Advances in Fatigue and Fracture Testing and Modelling

Edited by Zak Abdallah and Nada Aldoumani

Published in London, United Kingdom

IntechOpen

Supporting open minds since 2005

Advances in Fatigue and Fracture Testing and Modelling
http://dx.doi.org/10.5772/intechopen.94796
Edited by Zak Abdallah and Nada Aldoumani

Contributors
Haider Khazal, Nathera A. Saleh, Javad Sharifi Ghaderi, Gilberto Gomes, Thiago Oliveira, Francisco Evangelista Jr., Evren Meltem Toygar, Ahmet Gulakman, Barkallah Rachida, Rym Taktak, Jamel Bouaziz, Noamen Guermazi, Vidit Gaur, Sachin Kumar, Fahmi Zaïri

First published in London, United Kingdom, 2022 by IntechOpen
IntechOpen is the global imprint of INTECHOPEN LIMITED, registered in England and Wales, registration number: 11086078, 5 Princes Gate Court, London, SW7 2QJ, United Kingdom
Printed in Croatia

British Library Cataloguing-in-Publication Data
A catalogue record for this book is available from the British Library

Additional hard and PDF copies can be obtained from orders@intechopen.com

Advances in Fatigue and Fracture Testing and Modelling
Edited by Zak Abdallah and Nada Aldoumani
p. cm.
Print ISBN 978-1-83969-554-4
Online ISBN 978-1-83969-555-1
eBook (PDF) ISBN 978-1-83969-556-8

For EU product safety concerns:
IN TECH d.o.o., Prolaz Marije Krucifikse Kozulić 3, 51000 Rijeka, Croatia, info@intechopen.com or visit our website at intechopen.com.

Meet the editors

Dr. Zakaria Abdallah is the principal and a lead academic of fracture and fatigue in the Steel and Metals Institute, Swansea University, Wales. His role there involves in-house manufacturing (i.e., casting) of steel as well as testing its mechanical properties and fracture characteristics under various conditions. Dr. Abdallah has also worked at the Rolls-Royce University Technology Centre, Swansea University, where he was involved in the research and development of high-temperature alloys utilised in aero-engines. The power gearbox in the Rolls-Royce Ultrafan aero-engine is one of the major projects on which Dr. Abdallah has been working for several years. He has worked as a consultant for various industries in the United Kingdom, such as Airbus, TIMET, ETD, Rolls-Royce, within Swansea Materials Research and Testing (SMaRT) Ltd. Dr. Abdallah leads or has led, several modules at Swansea University. He also supervises several students working on various projects related to steel and metal alloys as well as the fatigue performance and life predictions of metallic structures. Dr. Abdallah has numerous journal publications, books, and international conferences to his credit. His research interests include steel and metals, composite materials, materials characterisation, creep and fatigue, life predictions of materials, thermo-mechanical testing, heat treatment, powder metallurgy, and diffusion bonding and brazing.

Dr. Aldoumani is the director of Innovative Technologies & Engineering Consultancy House (ITECH) Ltd., a registered business in the United Kingdom with headquarters in London. She provides services in metallic and composite structures in the field of uncertainty quantification and mechanical properties. In her business, Dr. Aldoumani is enhancing her consultancy expertise by providing services to customers worldwide. This includes finite element (FE) modelling and engineering solutions for newly designed products with the aid of uncertainty quantification and problem-solving. Dr. Aldoumani was a fellow researcher at Swansea University, Wales, with a structural engineering background. She is an expert in composite materials for structural applications, finite element (FE) modelling, and ANSYS, PDL, ACP, and MATLAB coding. Dr. Aldoumani is also a pioneer in uncertainty quantification in composite structures and adhesively bonded materials, particularly those employed in the aerospace industry. During her career, Dr. Aldoumani developed a novel approach that automates ANSYS through an in-house MATLAB code that is capable of running thousands of trials to obtain probabilistic Gaussian distributions. Her work is now published in internationally recognised and peer-reviewed journals.

Contents

Preface

Materials science is used to design and discover new materials for use in buildings, offices, cars, aircrafts, ships, rockets, and much more. Over the years, I have worked in all aspects of materials engineering, including development, alloying, casting, mechanical testing, and characterisation. I have always wanted to put my efforts into action by gathering researchers from around the world to contribute to the field and support other researchers to develop materials for various applications. This book is a result of those efforts. Over the years, various models have been proposed and utilised to predict the performance of materials under fatigue conditions. Some of these models proved to be accurate, whereas others failed to predict fatigue behaviour. The strength of any predictive tool depends on the employment of physically meaningful parameters that replicate real-life conditions. This book discusses fatigue testing and modelling in both metallic and non-metallic materials. It also presents predictive models and finite element (FE) activities to illustrate the behaviour of materials in real-life conditions. I am grateful to all the authors for their valuable efforts and contributions. I am also grateful to IntechOpen for publishing this important work.

Dr. Zak Abdallah
Steel and Metals Institute (SaMI),
Swansea University,
UK

Dr. Nada Aldoumani
Director - Innovative Technologies & Engineering
Consultancy House (ITECH) Ltd.,
UK

Chapter 1

Advances in Fatigue Prediction Techniques

Sachin Kumar and Vidit Gaur

Abstract

Fatigue is the prevalent mode of failure in engineering components made from metals. It occurs when the component is undergone fluctuating stresses. It leads to failure in metal structures because of damage accumulation. Therefore fatigue analysis of these structures is vital to ensure service reliability during the real operation. Among various simulation tools available to determine the fatigue life and mechanical behavior of metals and metallic components, the Crystal Plasticity Finite Element Method (CPFEM) has gained wide attention to analyze the microstructural attributes. Owing to the versatility of CPFEM in analyzing the fatigue properties of various metals and alloys, this chapter is aimed to examine and document the state-of-the-art research outcomes linked to fatigue behavior using CPFEM tool. The CPFEM is expected to accelerate the research progress to discover novel metals and alloys with better fatigue properties. For structural welds, estimating fatigue life depends on the material characteristics, object geometry, and applied loads. An in-depth analysis of stress concentrations across the affected areas of welds is vital for calculating fatigue response.

Keywords: crystal plasticity, fatigue life estimation, metallic structures, finite element method, weld sections

1. Introduction

It is well-documented that most of the failures in metallic components are because of fatigue. This makes it vital to analyze and understand the physics behind the fatigue failure and underline a relationship to minimize the chances of failure [1]. Estimation of material fatigue based on multiple experimentation and prediction can have a major role in safe and reliable mechanical design [2]. Several researchers have added their massive effort to devise sound and practical methodologies for fatigue prediction and assess mechanical structures' safety under cyclic mechanical loading [3–5]. It has been affirmed that an accurate forecast of fatigue life is complicated because one must consider several variables to avoid the catastrophic failure of engineering structures while in service [6]. The accuracy of fatigue prediction models is largely based on the capability to predict and model damage under non-zero superimposed mean stresses, range for multi-axiality across the stress regions, and concentration [7]. In case of cyclic and arbitrary loading, it is pretty cumbersome for predicting the fatigue properties because the damage is reliant on major stress components and their deviations while the loading [7].

2. Fatigue of welded sections

When we talk about the welded structure, the analysis of fatigue properties becomes even more critical. Whether solid-state or fusion [8–15], the welding processes can be considered the most widely used joining method to fabricate metallic structures, components, bridges, cranes, low and heavy-duty machines, etc. Several of these are designed and developed to have sustainability for a more extended period under fatigue loading. Typically in a welded component, the fatigue failure occurs either across root or toe of weld. Hence this becomes critical to align the design consideration to avoid premature failure under fatigue loading without incurring any noticeable cost. An initial assessment of the actual fatigue life may facilitate designers for obtaining objectives mentioned above.

Current industries put their massive effort into remote design, analysis, and validation of the structural/welded components, followed by their finite element analysis (FEM) to cater the future requirements [16]. This eliminates the time and cost involved in developing and test expensive pre-fabricated prototypes. The use of FEM also facilitates reducing cycle time and expediting inclusion to production sequence. This does also facilitates quick product development, faster launching into the external market. However, remote design and development of components face many challenges, some of which will be discussed in the forthcoming paragraphs.

As we know, that under fatigue loading, the structure or weld section undergoes permanent failure. The estimation of fatigue can be comprised of initiation of cracks and their propagation until final failure. The fatigue process is often found to originate across the stress concentration region, for example, across the weld toe. Nucleation cum spread of fatigue crack is measured based on the magnitude of stress across the favorable crack planes. Society of Automotive Engineers (SAE) has provided an easy information flow chart to determine and analyze stress and fatigue (**Figure 1**). Herein the structure geometry, material properties, and load profile are the significant variables to identify the fatigue life of a component. Usually, the weld joints are the weakest section in a structure, and fatigue failure is most likely to occur across these regions. Therefore to avoid the material failure under fatigue action, the designer needs to establish a relation to calculate the fatigue of structure and weld sections accurately.

In a review paper, Fricke [18] described various techniques to predict the fatigue life for seam welds. Niemi [19] and Fricke [20] covered various recommendations in context to analyze the fatigue behavior of weld joints. The importance

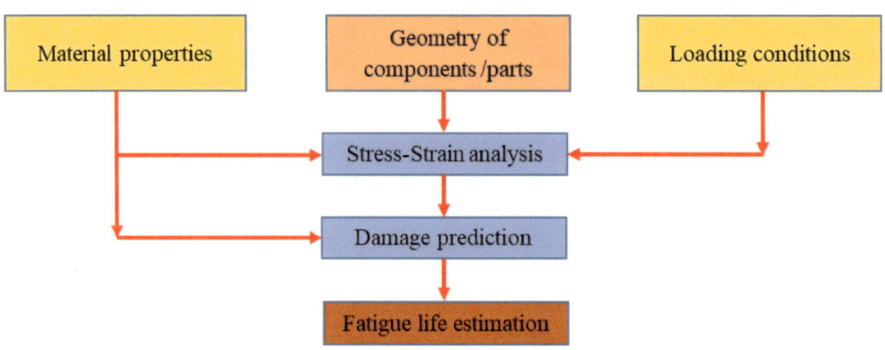

Figure 1.
Stages of stress and fatigue analysis [17].

of analyzing fatigue behavior of structures and weld sections can be understood by the fact that several national and international organizations such as the American Society of Mechanical Engineers (ASME), the International Institute of Welding (IIW), and the Society of Automotive Engineers (SAE), etc. are working round the clock to establish standard procedures and guidelines for determining the fatigue properties of weld joints. Various design guidelines and multiple updates have been proposed by the IIW on "recommendations for the fatigue design of welded components and structures". Various standard procedures are available for determining the fatigue properties of weld sections as per the assessment criterion and requirement of stress –strain data (**Figure 2**).

Previously, several models such as stress and strain-based models [21, 22], Critical plane models [23, 24], Enclosed surface models [25], and Integral type models [26], etc. have been used to predict fatigue properties of structural

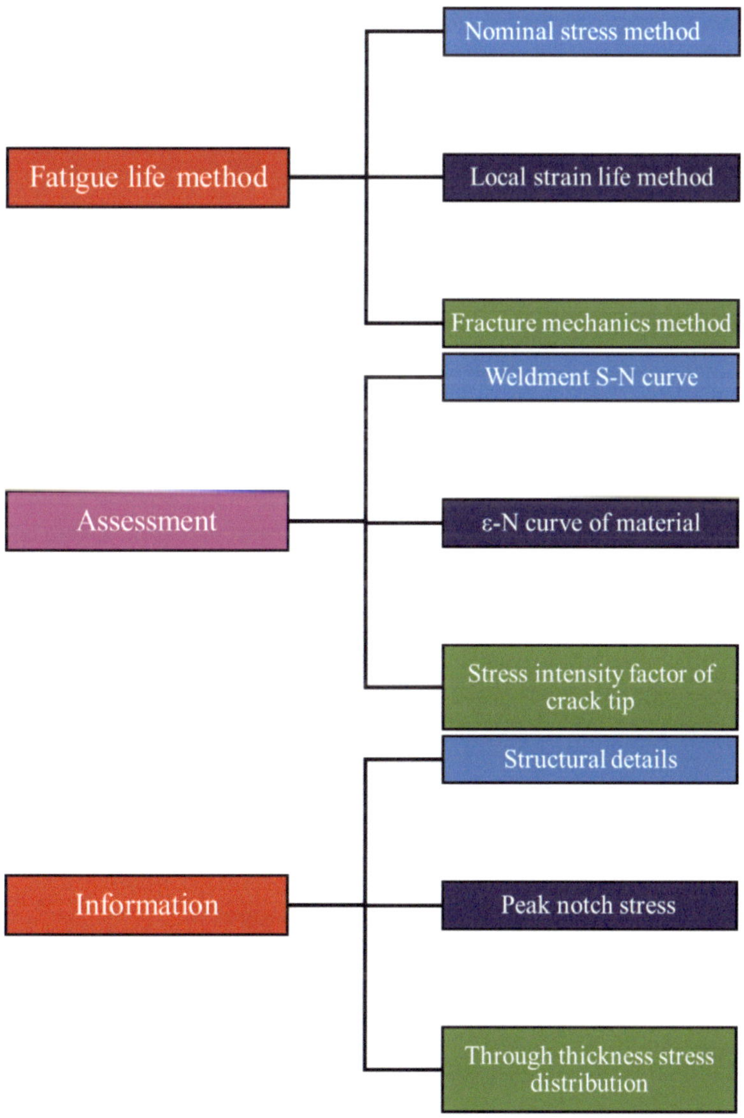

Figure 2.
Variables for fatigue life prediction.

components, however, could not estimate the fatigue properties accurately. The fatigue life predictions based on the application of empirical models (Basquin law, Miner's rule, Goodman diagram) taking into account the macroscopic mechanical fields are structure-oriented methods and therefore material-specific; thus, they do not account for the stochastic fatigue behavior due to the microstructural variabilities. Recently a methodology to predict the fatigue property of metal structures is gaining increasing acceptance and is known as CPFEM. The CPFEM can provide insights, and eventually, predictions of the fatigue behavior and variability of metallic materials become easier. CPFEM is also helpful to accelerate the development of new alloys with improved fatigue performances. While using the CPFEM, the advantage is that the fatigue predictions are much closer to the actual failure phenomenon and thus realistic. On the other hand, they require significant computational time and material intrinsic parameters that are difficult to measure experimentally. The accuracy of such simulations is strongly dependent on the synthetic microstructures generated through mathematical scripts or EBSD data. The CPFEM can provides insights and eventually predictions of the fatigue behavior and variability of metallic materials becomes easier. CPFEM is also helpful to accelerate the development of new alloys with improved fatigue performances.

3. Crystal Plasticity

Crystal plasticity (CP) is known as a fundamentally multi-dimensional approach beginning at the atomic level (dislocation cores) across sub-structural dislocation engagements in a single grain and up to the macroscopic mechanical behavior of the material. The multi-scale specialty and extensive research status of CP allow scientists and engineers to concentrate on different methodological domains. Plastic slip can be regarded as the utmost common plastic deformation mechanism in crystalline solids and metals. Typically, the slip system is found across the pre-established directions and planes in the form of plastic shearing, which is taken care of by the atomic arrangement at a regular crystal structure. The constitutive equations, theories, and mechanisms are derived keeping in mind the deformation characteristics of the materials on meso and micron levels. It derives the significant alteration of crystal plasticity from classical plasticity models for computational analysis. The hypothesis is derived much ahead in context to the continuum mechanics, stress fields, and other essential variables. For single and poly-crystals, the inelastic deformation models are successfully obtained and promoted the researchers to apply the crystal plasticity models for predicting the fatigue fracture and damage analysis. Besides, the freedom to apply various empirical formulas at the scale level permits solicitation of continuum dislocation simulations for determining crack initiation and propagation behavior [27].

4. Mechanism of dislocation slip: physical aspects

Crystals have different structural arrangements of atoms (**Figure 3**). The directions and planes are determined as per the Miller indices with reference to the unit cell edge vectors e_i. The slip plane m_g and direction s_g in FCC crystal are shown in **Figure 4**. Equivalent families of directions and planes for an FCC are depicted in **Figure 4**. For hexagonal crystals, four indices are often noticed to permit permutations when referring to families, one index along each possible edge direction, being the last one the height c direction, **Figure 3**. The FCC arrangement (**Figure 4a**) has

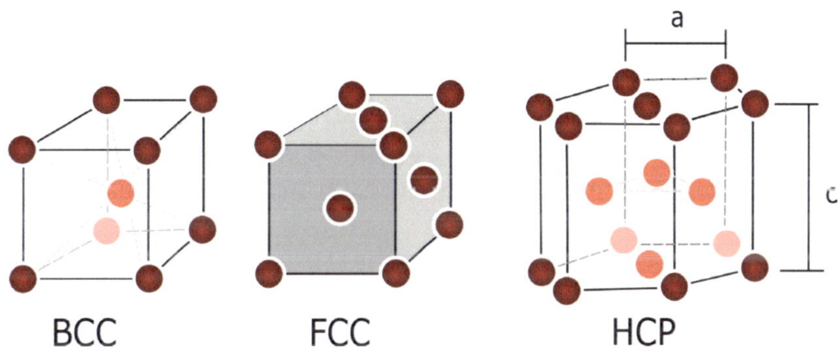

Figure 3.
BCC = body-Centered cubic. FCC = face-Centered cubic. HCP = hexagonal close-packed. The form ratio for HCP is c/a, ideally c/a = 1.632, but for example, cadmium has c/a = 1.886 and berilium has c/a = 1.586 [28, 29].

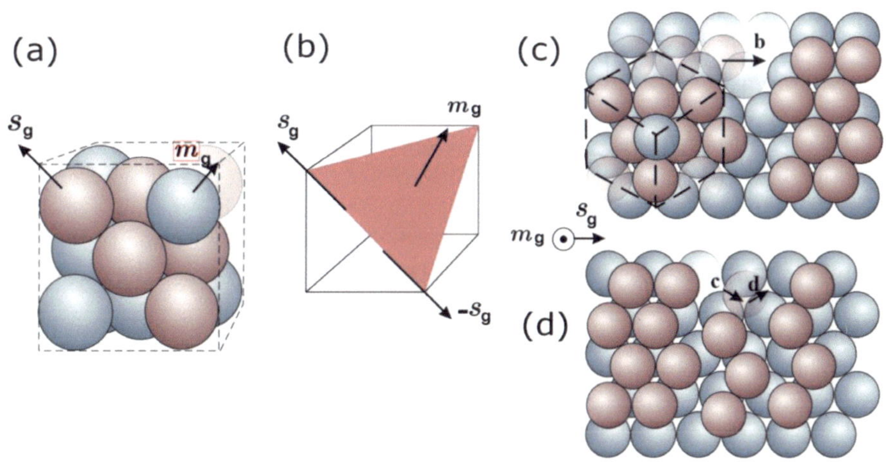

Figure 4.
(a, b) Slip plane m_g and direction s_g in a FCC crystal. (c) Total dislocation in a FCC. (d) Partial dislocation in a FCC lattice, where the dissociation is given as b = c + d = 1/6[$\bar{1}\,2\,\bar{1}$] + 1/6 [$\bar{2}\,11$]. Adapted from [28–30].

atoms at the vertices and center of the faces of an ideal hexahedron, resulting in a dense arrangement, and typical of especially ductile materials (e.g., Al, Cu, Au, Ag). FCC crystals have four {111} planes with three <110> directions in each plane, therefore in total have 12 slip systems, referred to as {111} <110>. Body-centered cubic crystals (BCC), (Cr, Fe, W) have atoms in the vertices and in the center of the hexahedron, and usual slip planes in the {110} family and direction < $\bar{1}$11> (secondary slip systems may become active for some materials and temperatures).

Dislocations are considered as crystal defects that were originated because of the missing of one or few atoms from the regular atomic hierarchy (**Figure 4c** and **d**). The slip is formed due to the sliding of several dislocations, eased by the stress field developed around them. This glide does not occur simultaneously across all the grain but takes place in a worm-like movement. This gliding emphasizes the movement of dislocation concerning a plane that may have both: Burger vector b and its line (**Figure 4c**). The dislocations and their associated local stress fields interact; blocks or auxiliary ease their movement. A slip band is formed because of the accumulation of these multiple displacements. The slip bands may be visualized across the regions of polished samples that might have undergone severe plastic deformation previously.

5. Crystal plasticity model

Even though multiple models of crystal elasto-plasticity and elasto-visco elastoplasticity have been added in several kinds of research to analyze the plastic distortion and fatigue fracture of various structures; there is an absence of in-depth and systematic documentation of the process mechanism involved and essential outcomes.

Nearly in the mid- 1980s, based on extensive experimentation and analysis, Panin et al. [31, 32] added a hypothesis to analyze a deformed material as a complex arrangement or multi-scale system. It proposed to originate a novel theme based on solid-state physics, also known as physical meso-mechanics. In other studies, Panin et al. [33, 34] documented the concepts involved and typical approaches for physical meso-mechanics. These studies were based on a classical assumption of considering a deformed solid as a multi-scale system where the plastic flow takes place because of a loss in shear stability in stress regions on variable atomic levels. These discussions (based on experimentation and theoretical hypothesis) resulted in a new approach to consider a deformed solid as a multi-scale self-organizing system. It permitted configuring a multi-scale model of a deformed object that may have a complete know-how of structural scales of deformation.

Firstly, Pierce et al. [35] proposed the concept of crystal plasticity finite element method (CPFEM) to study the tensile properties of a single crystal. They analyzed only two symmetric slip systems in their model at the initial stage because of the high computational cost. At later stage, Harren et al. [36] implemented the CPFEM to a polycrystalline material by applying a 2D model. They analyzed the mechanical characteristics for a polycrystalline Cu under tensile, compressive, and shear loading. The modeling of channel die compression in an FCC Al material under twelve slip systems is taken by Becker [37]. At a later stage, several other researchers used the CPFEM tool for various purposes due to its increased computational power. For nickel-base super-alloy, Manonukul and Dunne [38] added the CPFEM concept to analyze the crack formation characteristic in LCF and HCF. For the same material, Guan et al. [39] incorporated CPFEM and high-resolution digital image correlation technique to analyze the strain localization under fatigue. It is reported that for a wide range of parameters, the CPFEM can predict the material behavior very accurately.

Kysar [40, 41] analyzed the crack formation behavior across the copper/sapphire interface developed by diffusion welding by multiple experimentation and theoretical studies. They noticed that the propagation velocity of brittle crack was much higher compared to that of ductile one. The behavior of both the cracks was different, which may be due to the variation in slip system orientation in the case of single crystalline copper with respect to the direction of crack growth. In simple words, it can be said that variation in ductility may be due to the differences in dislocation substructure evolution that may promote variable stress properties at the fracture or crack tip. Van der Giessen et al. [42] reported an analysis of discrete dislocation simulation for material response at the crack tip. In their work, they took into account only the edge dislocations having dislocation lines that were perpendicular to the modeled plane. They observed that the stress values across the crack zone were significantly higher compared to those obtained in a direct elasto-plastic model. Flouriot et al. [43] proposed theoretical cum experimental analysis of stress rate in the close region of fatigue crack for FCC crystal. Their theoretical investigations were based on the crystal elasto-viscoplastic model [44] with power-law for shear rate. Across the crack region, strain localization bands, kink bands, and lattice rotation, the theoretical and experimental outcomes showed good agreement. To analyze the damage and fracture, Clayton [45] proposed two-level

DOI: http://dx.doi.org/10.5772/intechopen.99361

direct elasto-viscoplastic model. He detailed about the kinematic, dynamic, and thermodynamic properties as well as their relations at each level. In his analysis, an assumption to have internal discontinuities and displacement field discontinuities were made for kinematic equations. He incorporated crystal plasticity model added with anisotropic hardening law to take into account the effect of temperature field. Boudifa et al. [46] incorporated a self-consistent crystal plasticity model to investigate the fracture behavior and strain localization. To determine the damage in mesoscale, averaging of the required parameters over a representative macro volume was done.

For friction stir welding (FSW) [47–49] process, mechanical properties of joints are widely governed by microstructure patterns [50–52] and crystal structure characteristics. Across the stirred zone (SZ) in Al alloys, the strain field is not continuous for every texture band joint and contains high angle grain boundaries at most [53]. Some models have been employed to determine the influence of microstructure on the mechanical performance for the FSW joint. Dhondt et al. [53] used CPFEM to document the impact of texture on strain rate across the SZ of FSW joints. Romanova et al. [54] employed 3D microstructure-based model of FSW steel joints using a mesoscale deformation process. The effects of polycrystalline microstructure on the material flow process and fracture failure of FSW joints for Al alloy were studied by Balokhonov et al. [55]. In the case of FSW joints of Mg alloys, the texture analysis on deformation characteristics was modeled using a plastic finite element method by He et al. [56].

For laser-weld joints, Tu et al. [57] employed the Rousselier model for analyzing the fracture properties of Al alloys. Gaur et al. [58] added CPFEM technique to analyze the role of mean stress and weld defects to the fatigue life for Al alloys. They employed a two-dimensional model to simulate the fatigue loading, which replicated the microstructure pattern of the respective metal also. They added an anisotropic tessellation algorithm, as used by Briffod et al. [59] in their work for analyzing the grain shape and size taken by EBSD data. The orientation of crystals for each grain is governed by the algorithm proposed by Melchior and Delannay [60] and was relied on the orientation distribution function (ODF) of EBSD.

The R-ratio ($\sigma_{yy,\min}/\sigma_{yy,\max}$) distribution as an aggregate for the last loading cycle at the lowest applied stress ranges at the respective R-ratios is shown in **Figure 5**.

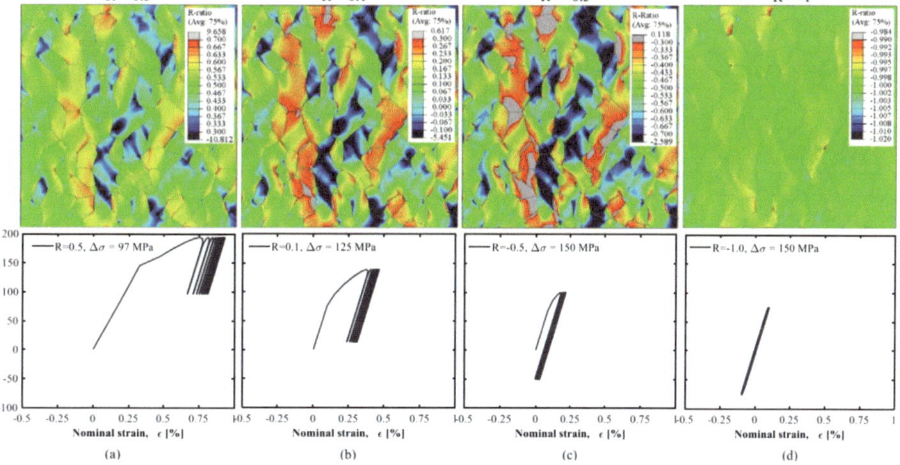

Figure 5.
Inhomogeneities in R- ratio distribution and macroscopic stress strain hysteresis curve at (a) Δσ = 97 MPa, R = 0.5 (b) Δσ = 125 MPa, R = 0.1 (c) Δσ = 150 MPa, R = -0.5 (d) Δσ = 150 MPa, R = -1 [58].

As depicted, a significant amount of heterogeneity for R-ratio distribution with respect to its increasing value is foreseen. When R-ratios are low, macroscopic and local R-ratios are more or less the same in most of the regions. Similarly, when R-ratios are low, their difference is high. A plot depicting the R-ratio distribution at the steady-state condition when $R = 0.5$, i.e., during the last load cycle, and maximum and minimum stress values are shown in **Figure 6**. For a fair interpretation, a plot of stress vs. strain is made across the four randomly chosen elements. In the case of far-field, R-ratio comes out 0.5; however, R-ratio across the four elements are 0.54, −0.61, 5.37 and −4.32. The results emphasized that precise prediction of crack initiation could not be possible by macroscopic parameters. It is because some inhomogeneities may be developed at the microstructural level. This outcome backs the results reported in the previously published literature [61]. While using the MIG technique to join Al-Mg alloy with different filler-wire, Gaur et al. [62, 63] analyzed the fatigue properties of weld joints. It was observed that the damaging action of mean stress was because of a decrease in crack-nucleation time and crack closure effects. Also, at low R-ratios (<0.1), maximum fatigue failures were surface-initiated. However, at high R-ratios ($\geq \mathbf{0.1}$), mostly the defect-induced failures were predominant. A phenomenon of shift in crack development mechanism can be understood because of local cyclic plasticity under stress-concentration factor.

Several researchers [58, 64, 65] have implemented the CPFEM model to predict the material properties and found that the predicted results were in good agreement with the experimental data, of course within some acceptable scatter. For example, Ye et al. [65] analyzed the fatigue crack initiation behavior in an Al beam with a hole under 4-point bending. They employed both in-situ experiments (EBSD and digital image correlation) and CPFEM simulations to investigate the slip bands and crack initiation sites at the microstructure scale. Based on EBSD maps, a realistic microstructure model was developed. They noted that the simulation results had a good agreement with the experimental outcomes in several aspects. Gaur et al. [58] predicted fatigue lives and its comparison with experimental data at different R-ratios without considering any defect (**Figure 7**). The anticipated results are observed to have a good agreement with the experimental outcomes for all the R-ratios. It is also

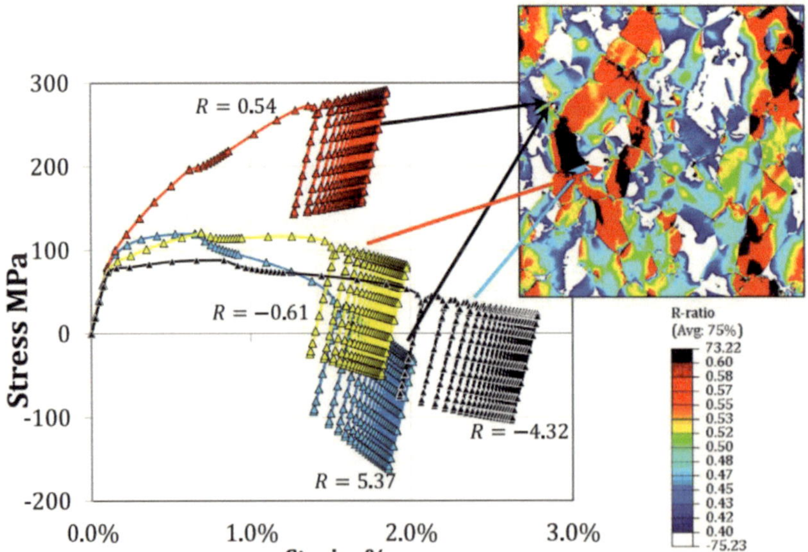

Figure 6.
An example of evolution of stress–strain loops at different locations in an aggregate at $\Delta\sigma = 135$, $R = 0.5$ [58].

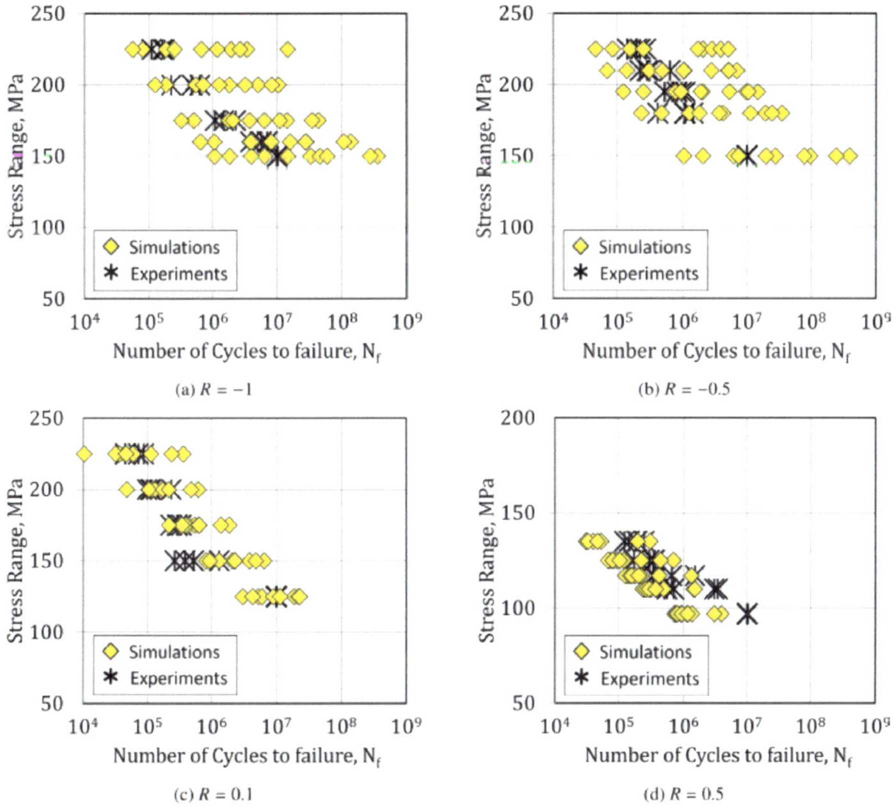

Figure 7.
Prediction of fatigue lives at different R-ratios without considering any defect [58].

important to note that the scatter in fatigue lives has increased significantly upon considering defects in FE simulations.

6. Conclusions

This chapter has documented a systematic review of previous works on fatigue and fracture analysis by considering the CP approach of weld joints. As per the findings, several models based on the CPFEM methodology have standard features and details the micro-damage initiation and transition to failure. As indicated in numerous articles, the hardening model has a significant role in influencing computational outcomes. It is because hardening law can be directly related to depicting the defect formation mechanism in materials and that "transforms" to the micro-damage structure. The anisotropy of crystal can largely influence the micro-damage formation system to the extent of each grain and by grain lattice rotation adjacent to the crack tip region.

Several new approaches in context to fatigue analysis are reported in the literature duly supported with pertinent details. It is seen that the proposed models have their limitations and certain conditions, i.e., low cycle fatigue or high cycle fatigue, uniaxial and multiaxial loading, and material type (ductile or brittle), etc. Unfortunately, no model is found to fit perfectly under variable loading and material conditions. It leaves a wide gap for further research and development to the precise and reliable prediction of fatigue life under different conditions. The chapter was intended to present a short review to explore novel concepts and techniques, such as, CPFEM, to analyze and predict the fatigue life of weld joints, unlike some classical approaches, which lack to have a generalized algorithm to model the fatigue life. Future research should be oriented to the implementation of optimization models/algorithms in order to explore their full capability for accurately predicting the fatigue properties of metal structures or welded joints.

Author details

Sachin Kumar[1] and Vidit Gaur[2]*

1 Department of Mechanical Engineering, Indian Institute of Science (IISc), Bengaluru, Karnataka, India

2 Department of Mechanical and Industrial Engineering, Indian Institute of Technology (IIT), Roorkee, UK, India

*Address all correspondence to: vidit.gaur@me.iitr.ac.in

References

[1] Sangid MD. The physics of fatigue crack initiation. Int J Fatigue 2013;57:58-72. https://doi.org/10.1016/j.ijfatigue.2012.10.009.

[2] Lee Y-L, Pan J, Hathaway R, Barkey M. Fatigue Testing and Analysis: Theory and Practice. New York Elsevier Butterworth Heinemann 2005. https://books.google.co.in/books?hl=en&lr=&id=mrPFTVr2hH4C&oi=fnd&pg=PP2&ots=nOW4dbH6sg&sig=zPqvmLkfw6D43RIy7nRKnS8cAf4&redir_esc=y#v=onepage&q&f=false (accessed May 18, 2021).

[3] Susmel L, Taylor D. A critical distance/plane method to estimate finite life of notched components under variable amplitude uniaxial/multiaxial fatigue loading. Int J Fatigue 2012;38:7-24. https://doi.org/10.1016/j.ijfatigue.2011.11.015.

[4] Rahman MM, Ariffin AK, Abdullah S, Jamaludin N. Finite Element Based Durability Assessment of a Free Piston Linear Engine Component. vol. 3. 2007.

[5] Rahman MM, Ariffin AK, Rejab MRM, Kadirgama K, Noor MM. Multiaxial fatigue behavior of cylinder head for a free piston linear engine. J Appl Sci 2009;9:2725-2734. https://doi.org/10.3923/jas.2009.2725.2734.

[6] Brighenti R, Carpinteri A. A notch multiaxial-fatigue approach based on damage mechanics. Int J Fatigue 2012;39:122-133. https://doi.org/10.1016/j.ijfatigue.2011.02.003.

[7] Susmel L, Taylor D. The modified Wöhler curve method applied along with the theory of critical distances to estimate finite life of notched components subjected to complex multiaxial loading paths. Fatigue Fract Eng Mater Struct 2008;31:1047-1064. https://doi.org/10.1111/j.1460-2695.2008.01296.x.

[8] Kumar S, Ding W, Sun Z, Wu CS. Analysis of the dynamic performance of a complex ultrasonic horn for application in friction stir welding. Int J Adv Manuf Technol 2018;97:1269-1284. https://doi.org/10.1007/s00170-018-2003-0.

[9] Kumar S, Wu CS. A novel technique to join Al and Mg alloys: Ultrasonic vibration assisted linear friction stir welding. Mater Today Proc 2018;5:18142-18151. https://doi.org/10.1016/j.matpr.2018.06.150.

[10] Kumar S. Ultrasonic assisted friction stir processing of 6063 aluminum alloy. Arch Civ Mech Eng 2016;16:473-484. https://doi.org/10.1016/j.acme.2016.03.002.

[11] Kumar S, Wu C. Strengthening Effects of Tool-Mounted Ultrasonic Vibrations during Friction Stir Lap Welding of Al and Mg Alloys. Metall Mater Trans A Phys Metall Mater Sci 2021. https://doi.org/10.1007/s11661-021-06282-w.

[12] Gaur V, Doquet V, Persent E, Roguet E. Corrosion-fatigue behaviour of Cr–Mo steel under biaxial tension. Fatigue Fract Eng Mater Struct 2020;43:2560-2570. https://doi.org/10.1111/ffe.13276.

[13] Yadav VK, Gaur V, Singh I V. Effect of post-weld heat treatment on mechanical properties and fatigue crack growth rate in welded AA-2024. Mater Sci Eng A 2020;779:139116. https://doi.org/10.1016/j.msea.2020.139116.

[14] Gaur V, Doquet V, Persent E, Roguet E. Effect of biaxial cyclic tension on the fatigue life and damage mechanisms of Cr-Mo steel. Int J Fatigue 2016;87:124-131. https://doi.org/10.1016/j.ijfatigue.2016.01.021.

[15] Gaur V, Doquet V, Persent E, Roguet E. Fatigue of Clip Connectors for

Offshore Drilling Risers under the Combined Influence of High Mean Stress and Biaxial Tension. Procedia Eng., vol. 133, Elsevier Ltd; 2015, p. 90-101. https://doi.org/10.1016/j.proeng.2015.12.631.

[16] Kumar S, Wu CS. Review: Mg and Its Alloy - Scope, Future Perspectives and Recent Advancements in Welding and Processing. J Harbin Inst Technol 2017;24:1-37.

[17] Glinka G. Fatigue & Fracture Mechanics - Analysis & Design. Univ Waterloo, Lect Notes 2011.

[18] Fricke W. Fatigue analysis of welded joints: State of development. Mar Struct 2003;16:185-200. https://doi.org/10.1016/S0951-8339(02)00075-8.

[19] Niemi E. Recommendations concerning stress determination for fatigue analysis of welded components. 1995.

[20] Fricke W. Recommended Hot-Spot Analysis Procedure For Structural Details of Ships And FPSOs Based On Round-Robin FE Analyses | International Journal of Offshore and Polar Engineering | OnePetro. Int J Offshore Polar Engg 2002;12:40-48.

[21] Iseki T, Kameda T, Maruyama T. Interfacial reactions between SiC and aluminium during joining. J Mater Sci 1984;19:1692-1698. https://doi.org/10.1007/BF00563067.

[22] Findley WN. A Theory for the Effect of Mean Stress on Fatigue of Metals Under Combined Torsion and Axial Load or Bending. J Eng Ind 1959;81:301-305. https://doi.org/10.1115/1.4008327.

[23] Kamal M, Rahman MM. Advances in fatigue life modeling: A review. Renew Sustain Energy Rev 2018;82:940-949. https://doi.org/10.1016/j.rser.2017.09.047.

[24] Susmel L, Lazzarin P. A bi-parametric Wöhler curve for high cycle multiaxial fatigue assessment. Fatigue Fract Eng Mater Struct 2002;25:63-78. https://doi.org/10.1046/j.1460-2695.2002.00462.x.

[25] Gonçalves CA, Araújo JA, Mamiya EN. Multiaxial fatigue: A stress based criterion for hard metals. Int J Fatigue 2005;27:177-187. https://doi.org/10.1016/j.ijfatigue.2004.05.006.

[26] Papadopoulos I V. A new criterion of fatigue strength for out-of-phase bending and torsion of hard metals. Int J Fatigue 1994;16:377-384. https://doi.org/10.1016/0142-1123(94)90449-9.

[27] Trusov P V., Ashikhmin VN, Volegov PS, Shveykin AI. Constitutive relations and their application to the description of microstructure evolution. Phys Mesomech 2010;13:38-46. https://doi.org/10.1016/j.physme.2010.03.005.

[28] Nguyen K, Zhang M, Amores VJ, Sanz MA, Montáns FJ. Computational modeling of dislocation slip mechanisms in crystal plasticity: A short review. Crystals 2021;11:1-29. https://doi.org/10.3390/cryst11010042.

[29] Abbaschian R, Reed-Hill RE. Physical Metallurgy Principles - SI Version. Cengage Learning: Stamford, CT, USA, 2009; 2009.

[30] Borja RI. Plasticity: Modeling & Computation. Springer Science & Business Media; 2013.

[31] Panin VE, Grinyaev Y V., Elsukova TF, Ivanchin AG. Structural levels of deformation in solids. Sov Phys J 1982;25:479-497. https://doi.org/10.1007/BF00898745.

[32] Panin VE, Grinyaev YV, Danilov VI, Zuev LB. Structural levels of plastic deformation and fracture. 1990.

[33] Panin VE. Methodology of physical mesomechanics as a basis for model

construction in computer-aided design of materials. Russ Phys J 1995;38:1117-1131. https://doi.org/10.1007/BF00559394.

[34] Panin VE, Grinyaev Y V. Physical mesomechanics: a new paradigm at the interface of solid state physics and solid mechanics. Phys Mesomech 2003;4:7-32.

[35] Peirce D, Asaro RJ, Needleman A. An analysis of nonuniform and localized deformation in ductile single crystals. Acta Metall 1982;30:1087-1119. https://doi.org/10.1016/0001-6160(82)90005-0.

[36] Harren S V., Dève HE, Asaro RJ. Shear band formation in plane strain compression. Acta Metall 1988;36:2435-2480. https://doi.org/10.1016/0001-6160(88)90193-9.

[37] Becker R. Analysis of texture evolution in channel die compression-I. Effects of grain interaction. Acta Metall Mater 1991;39:1211-1230. https://doi.org/10.1016/0956-7151(91)90209-J.

[38] Manonukul A, Dunne FPE. High- and low-cycle fatigue crack initiation using polycrystal plasticity. Proc R Soc A Math Phys Eng Sci 2004;460:1881-1903. https://doi.org/10.1098/rspa.2003.1258.

[39] Guan Z, Liu J. Sudden occurence of hypotension and bradycardia during greenlight laser transurethral resection of prostate: Case report of two cases. BMC Anesthesiol 2016;16:1-4. https://doi.org/10.1186/s12871-016-0234-x.

[40] Kysar JW. Continuum simulations of directional dependence of crack growth along a copper/sapphire bicrystal interface. Part II: Crack tip stress/deformation analysis. J Mech Phys Solids 2001;49:1129-1153. https://doi.org/10.1016/S0022-5096(00)00071-5.

[41] Kysar JW. Continuum simulations of directional dependence of crack growth along a copper/sapphire bicrystal interface. Part I: Experiments and crystal plasticity background. J Mech Phys Solids 2001;49:1099-1128. https://doi.org/10.1016/S0022-5096(00)00072-7.

[42] Van Der Giessen E, Deshpande VS, Cleveringa HHM, Needleman A. Discrete dislocation plasticity and crack tip fields in single crystals. J Mech Phys Solids 2001;49:2133-2153. https://doi.org/10.1016/S0022-5096(01)00040-0.

[43] Flouriot S, Forest S, Cailletaud G, Köster A, Rémy L, Burgardt B, et al. Strain localization at the crack tip in single crystal CT specimens under monotonous loading: 3D Finite Element analyses and application to nickel-base superalloys. Int J Fract 2003;124:43-77. https://doi.org/10.1023/B:FRAC.0000009300.70477.ba.

[44] Trusov P V., Shveykin AI. Multilevel crystal plasticity models of single- and polycrystals. Statistical models. Phys Mesomech 2013;16:23-33. https://doi.org/10.1134/S1029959913010037.

[45] Clayton JD. Continuum multiscale modeling of finite deformation plasticity and anisotropic damage in polycrystals. Theor Appl Fract Mech 2006;45:163-185. https://doi.org/10.1016/j.tafmec.2006.03.001.

[46] Boudifa M, Saanouni K, Chaboche JL. A micromechanical model for inelastic ductile damage prediction in polycrystalline metals for metal forming. Int J Mech Sci 2009;51:453-464. https://doi.org/10.1016/j.ijmecsci.2009.03.014.

[47] Kumar S, Wu CS, Shi L. Intermetallic Diminution during Friction Stir Welding of Dissimilar Al/Mg Alloys in Lap Configuration via Ultrasonic Assistance. Metall Mater Trans A 2020;51:5725-5742.

[48] Kumar S, Wu CS. Suppression of intermetallic reaction layer by ultrasonic assistance during friction stir welding of

Al and Mg based alloys. J Alloys Compd 2020;827:154343. https://doi.org/https://doi.org/10.1016/j.jallcom.2020.154343.

[49] Kumar S, Wu CS, Song G. Process parametric dependency of axial downward force and macro- and microstructural morphologies in ultrasonically assisted friction stir welding of Al/ Mg alloys. Metall Mater Trans A 2020;51:2863-2881. https://doi.org/https://doi.org/10.1007/s11661-020-05716-1.

[50] Sato Y, Kokawa H, Enomoto M, Jogan S. Microstructural evolution of 6063 aluminum during friction-stir welding. Metall Mater 1999.

[51] Kumar S, Wu CS, Sun Z, Ding W. Effect of ultrasonic vibration on welding load, macrostructure, and mechanical properties of Al/Mg alloy joints fabricated by friction stir lap welding. Int J Adv Manuf Technol 2019;100:1787-1799. https://doi.org/10.1007/s00170-018-2717-z.

[52] Kumar S, Wu CS, Padhy GK, Ding W. Application of ultrasonic vibrations in welding and metal processing: A status review. J Manuf Process 2017;26:295-322.

[53] Dhondt M, Aubert I, Saintier N, Olive JM. Mechanical behavior of periodical microstructure induced by friction stir welding on Al-Cu-Li 2050 alloy. Mater Sci Eng A 2015;644:69-75. https://doi.org/10.1016/j.msea.2015.05.072.

[54] Romanova V, Balokhonov R, Schmauder S. Three-dimensional analysis of mesoscale deformation phenomena in welded low-carbon steel. Mater Sci Eng A 2011;528:5271-5277. https://doi.org/10.1016/j.msea.2011.03.065.

[55] Balokhonov RR, Romanova VA, Martynov SA, Zinoviev A V., Zinovieva OS, Batukhtina EE. A computational study of the microstructural effect on the deformation and fracture of friction stir welded aluminum. Comput Mater Sci 2016;116:2-10. https://doi.org/10.1016/j.commatsci.2015.10.005.

[56] He W, Zheng L, Xin R, Liu Q. Microstructure-based modeling of tensile deformation of a friction stir welded AZ31 Mg alloy. Mater Sci Eng A 2017;687:63-72. https://doi.org/10.1016/j.msea.2017.01.053.

[57] Tu HY, Schmauder S, Weber U. Numerical Simulation of Crack Propagation in Al6061 Laser Welded Joints. Procedia Mater Sci 2014;3:414-20. https://doi.org/10.1016/J.MSPRO.2014.06.069.

[58] Gaur V, Briffod F, Enoki M. Micro-mechanical investigation of fatigue behavior of Al alloys containing surface/superficial defects. Mater Sci Eng A 2020;775:138958. https://doi.org/10.1016/j.msea.2020.138958.

[59] Briffod F, Shiraiwa T, Enoki M. Microstructure modeling and crystal plasticity simulations for the evaluation of fatigue crack initiation in α-iron specimen including an elliptic defect. Mater Sci Eng A 2017;695:165-177. https://doi.org/10.1016/j.msea.2017.04.030.

[60] Melchior MA, Delannay L. A texture discretization technique adapted to polycrystalline aggregates with non-uniform grain size. Comput Mater Sci 2006;37:557-564. https://doi.org/10.1016/j.commatsci.2005.12.002.

[61] Gaur V, Doquet V, Persent E, Mareau C, Roguet E, Kittel J. Surface versus internal fatigue crack initiation in steel: Influence of mean stress. Int J Fatigue 2015;82:437-448. https://doi.org/10.1016/j.ijfatigue.2015.08.028.

[62] Gaur V, Enoki M, Okada T, Yomogida S. A study on fatigue behavior

of MIG-welded Al-Mg alloy with different filler-wire materials under mean stress. Int J Fatigue 2018;107: 119-129. https://doi.org/10.1016/j.ijfatigue.2017.11.001.

[63] Gaur V, Enoki M, Yomogida S. Physically short and long-crack growth behavior of MIG welded Al-5.8%Mg alloy. Eng Fract Mech 2019;209:301-316. https://doi.org/10.1016/j.engfracmech.2019.01.026.

[64] Adzima F, Balan T, Manach PY, Bonnet N, Tabourot L. Crystal plasticity and phenomenological approaches for the simulation of deformation behavior in thin copper alloy sheets. Int J Plast 2017;94:171-191. https://doi.org/10.1016/j.ijplas.2016.06.003.

[65] Ye W, Efthymiadis P, Pinna C, Ma A, Shollock B, Dashwood R. Experimental and modelling study of fatigue crack initiation in an aluminium beam with a hole under 4-point bending. Int J Solids Struct 2018;138:87-96. https://doi.org/10.1016/j.ijsolstr.2018.01.001.

Chapter 2

Failure Modes in Fiber Reinforced Composites and Fracture Toughness Testing of FRP

Evren Meltem Toygar and Ahmet Gulakman

Abstract

In this paper, interlaminar fracture behavior of woven-fabric-reinforced glass/epoxy composites has been investigated experimentally and numerically. The mechanical properties of this composite were studied and Mode I (Tensile Opening) DCB (Double Cantilever Beam) tests were performed on Fiber Reinforced Composite (FRP) specimens to determine the delaminating resistance of composite laminates used for structural applications. Techniques for measuring the interlaminar fracture toughness, K_{IC} data of woven-fabric-reinforced glass/epoxy composite materials, are highlighted under the consideration of ASTM Standard D5528–01 and test methods ISO 15024, DIN EN ISO 75-1 and DIN EN ISO 75-3. The obtained test results were apparently consistent with the assumptions of the CCM (Compliance Calibration Method) that was used to obtain the interlaminar critical SERR (strain energy release rate), G_{IC}. Finite element analysis was conducted to validate the closed form solution. The use of obtained mechanical properties data in finite element analyses utilizing fracture mechanics are examined. Results show a good agreement between experimental and numerical solutions.

Keywords: interlaminar fracture toughness, fracture mechanics, woven-fabric-reinforced glass/epoxy composites, DCB test method, finite element method

1. Introduction

Woven fabric reinforced composites are the most important and widely used forms among textile structural composites. In recent years, fracture mechanics has found an extensive applications in damage analysis of composite laminates, especially in delamination analysis. Delaminations may occur during manufacture because of incomplete curing or may result from impact damage; or they may result from the interlaminar stresses that develop at stress-free edges or discontinuities. It has been mentioned that delamination in a composite laminate usually occurs at the interface of different oriented plies and tends to grow and also it can be a major problem for laminated composite structures. Sometimes, delaminations may also be result of contamination of the pre-preg, the poor ply adhesion, or they may form locally in regions of high void content. The problems of interlaminar performance are discussed along with the technique used to measure them and the fracture mechanics principles applied to improve them, because laminated composite materials have been increasingly used in large transport aircraft structures due to their low density compared to other materials. The delamination resistance of laminated composites can be measured by critical SERR (strain energy release rates). The DCB (Double Cantilever Beam) is the most popular specimen configurations in the experimental determination of Mode I interlaminar fracture toughness. Aliyu and Daniel [1], used conventional double cantilever beam specimens with different configurations to measure the Mode I fracture toughness in a graphite/epoxy composite. The effect of stacking sequence on energy release rate distribution across the specimen width the multidirectional DCB specimens are determined by Davidson et al. [2]. They investigated eight different stacking sequences, with the delamination growth between 30°/−30° interface.

Sun and Zheng [3], analyzed the distribution of strain energy release rate, G at the crack fronts of double cantilever beam specimens by means of the plate finite element. They found a boundary layer phenomenon in the distribution of G at the crack front, which causes the strain energy release rate to vary along the straight crack front. Toygar et al. [4] measured the fracture toughness value of carbon/epoxy composite materials by using the CMOD (crack mouth opening displacement) method experimentally using SENT (Single Edge Notch Tension) specimens. The finite element study was carried out by using 2-D model to obtain the fracture toughness value of woven carbon/epoxy composite numerically with the ABAQUS finite element software package. Mode I DCB tests were performed on carbon/epoxy woven laminates in weave style, which means as plain weave such as $[0°/90°]_{16}$, where 16 means the number of layers, lay-up specimens by Morais et al. [5]. The starter crack was created at mid-thickness, between the 0° and 90°, stand for the directions of fill and warp strips, respectively. The test results were apparently consistent with the assumptions of the CBT (Corrected Beam Theory) that was used to obtain the interlaminar critical SERR, G_{IC}. The measured values were higher than those of unidirectional $[0°]_{24}$ specimens, especially the final propagation values. A finite-element analysis confirmed the applicability of the CBT for interlaminar propagation along the two [0°/90°] interfaces. They have found that the intralaminar G_{IC} is significantly smaller than the interlaminar G_{IC}. This will prevent pure interlaminar propagation in multi-directional specimens with high interlaminar fracture toughness. The interlaminar fracture is a common mode of failure for composite materials, especially in laminated architectures [6–8]. Interlaminar fracture mechanics has proven useful for characterizing the onset and growth of delaminations. Delamination onset and debonding in fracture toughness specimens (such as Double Cantilever Beam Specimen) and laboratory size coupon type specimens (such as the skin/stringer debond specimen) has been investigated [9–11].

In this study, the influence of test parameters on Mode I delamination resistance is examined. A detailed discussion on test method and data reduction schemes for Mode I interlaminar fracture toughness characterization of composite laminates have been done by considering the test parameters such as width, different fiber orientation and lay-up and ext. In the case of laminated composites, the Griffith criterion is applied to calculate the critical amount of energy required to propagate a crack and DCB test is used to obtain this material property. In a result of this, DCB tests on woven-fabric-reinforced glass/epoxy specimens were performed in accordance with ASTM Standard D5528–01 and ISO 15024, DIN EN ISO 75-1 and DIN EN ISO 75-3 Standards [12–16] to measure the delamination resistance under Mode I loading. Therefore, the present research has been focused on the characterization of delamination growth in laminated structures and has been done to determine Mode I interlaminar fracture toughness, K_{IC} of woven laminates $[0°/90°]_{16}$ and $[\pm 45°]_{16}$ lay-up composites such as in weave style as plain weave and with 16 number of layers. The obtained experimental data were reduced by using data reduction technique CCM (Compliance calibration method) to determine the critical SERR, G_{IC}. CCM generates a least squares plot of log ($\delta i/Pi$) versus log (ai), where i represents the number of specimen and changes from 1 to minimum 5 and P and δ represent the measured values of load and load point displacement, respectively during the test, and a represents the crack length, to use the visually observed delamination onset values and all the propagation values [17]. Additionally, fracture mechanics based finite element models of described DCB tests were developed to confirm the experimental results. The finite element analysis has been carried out to accomplish the delamination analysis of the specimen. Critical load levels, the geometrical and material properties of the test specimens were used as input data for the analysis to evaluate the Mode I energy release rate at the onset of delamination crack propagation.

2. Materials, sample preparation and measurement of mechanical properties

Delamination growth has typically been related to the cyclic SERR, G, for composite materials, using a power law expression [18–20], known as the Paris Law. Fracture toughness may be described as the critical energy that a material may absorb before failure. Independent of specimen geometry, this property describes the general resistance of a material to delaminate [21].

In this study, DCB specimens having density 1,72 g/cm^3 were manufactured in Dokuz Eylul University laboratory from the woven-fabric-reinforced glass/epoxy materials and the detail of sample cutting $[0°/90°]_{16}$ and $[\pm 45°]_{16}$ woven DCB specimens are given in **Figure 1a** and **b**, respectively. All were prepared from manufactured woven composite plate. The prepared composite plate has sixteen laminas and has a volume fraction of glass fiber of approximately 60%. During manufacturing, a thin non-adhesive Teflon film with 0.07 mm thickness were placed at the mid-plane (between 8th and 9th plies) at one end to simulate an initial delamination by considering ASTM Standard D5528–01 and British test method ISO 15024.

The mechanical tests to obtain the mechanical properties were carried out by using the prepared specimen which is given in **Figure 2**. The universal mechanical test machine, which is used, is available with different clamping techniques and accessories such as digital click-on extensometer from Zwick Roell. The width and thickness of each specimen were measured using a micrometer, at the center and each end. The tensile test was conducted following ASTM Standard E-681 [22].

2.1 Video extensometer

Video extensometer ME-46 is a high resolution and non-contact strain measurement system which is capable of measuring both longitudinal and lateral strain, modulus and Poisson ratio of polymer composites (see in **Figure 3a**). The advantage of the non-contacting method is that the strain is directly measured and end-effects from the gripping system need not to be considered [24] and for non-contact strain measurement the detail is given in **Figure 3b**. The measurement was carried out optically and there is no requirement for physical contact between the extensometer and the test specimen. Therefore, it can be applied to all sample without and slippage and related test specimen breakage problems [23]. The mechanical properties of this woven-fabric-reinforced glass/epoxy composite were obtained experimentally by using video extensometer under the consideration of ASTM Standards and were given in **Table 1**.

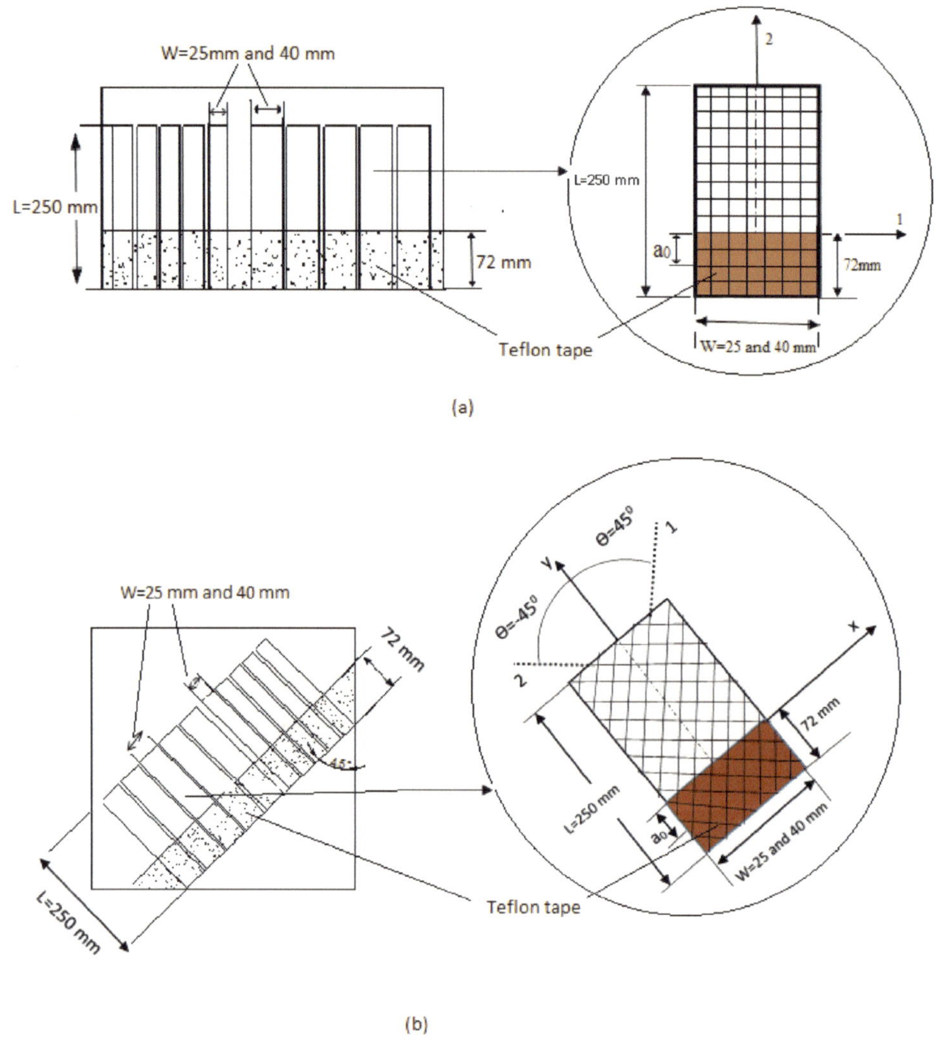

Figure 1.
The detail of cutting of sample from manufactured woven-fabric-reinforced glass/epoxy composite plate a) (0°/90°) fiber orientation, b) (±45°) fiber orientation.

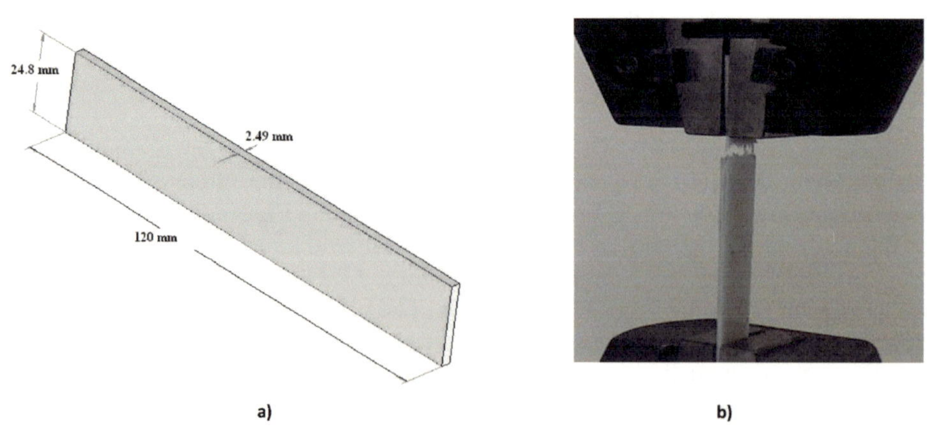

Figure 2.
a) Geometries of test specimen for mechanical properties testing, b) view of tension test.

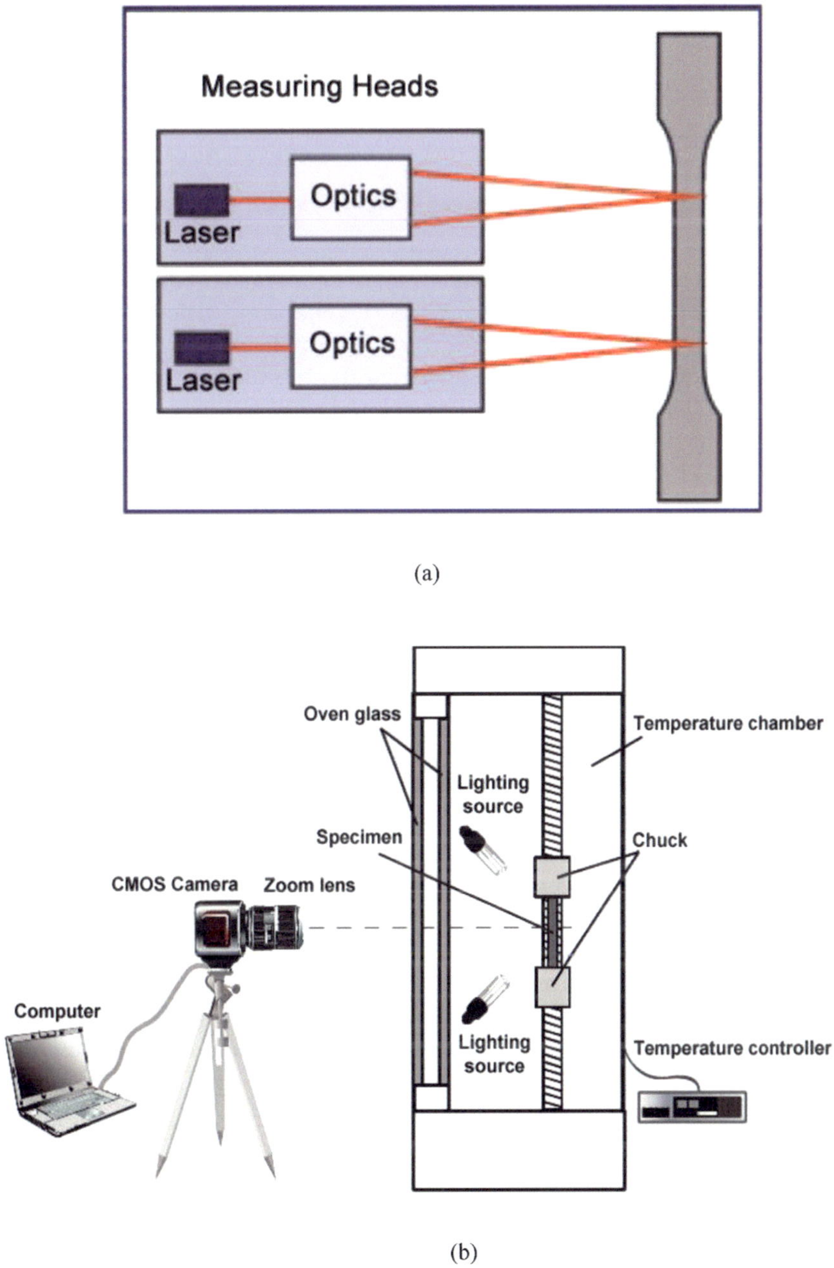

Figure 3.
a) Schematic detail of video extensometer [23], b) schematic drawing of the non-contact varying temperature deformation measuring system [24].

2.2 Optical deformation and strain measurement system

Aramis [21] is an optical technique from GOM (Gesellschaft fur Opticsche Mestechnik), Germany. It can be used to non-destructively measure the deformation and strain profiles of an object surface under loading. It has advantages of simple specimen preparation, large measuring area, non-contact and full field measurement, no laser illumination, material independent determination, full field and graphical results, three-dimensional results and good mobility. Its capabilities

Materials	Density (g/cm^3)	Poisson ratio (ν)		Modulus of elastisity (E) (Gpa)		Shear modulus (G) (Gpa)
		ν_{12}	ν_{21}	E_1	E_2	G_{12}
$(0°/90°)_{16}$	1.72	0.28	0.28	28.2	28.2	4.7

Table 1.
Physical and mechanical properties of woven specimen in the principal material direction.

include materials testing, stability estimating, components dimensioning, nonlinear behavior examination, creep and aging processes characterization. Because the deformation and strain of an object under mechanical loading is associated with its structure integrity, the abnormalities and irregularities in deformation and strain profiles indicate the damage presence. It can be adapted to assess the structural integrity and non-destructive testing of polymer composites.

2.3 DCB specimen for interlaminar fracture toughness

The average specimen width, b, was 25 mm and/or 40 mm for DCB specimen and an illustration of it was given in **Figure 4**. The hinges were mounted on the top and bottom surfaces of the end of DCB specimen arms by using an epoxy adhesive, which was cured. The average thickness h, was 2.5 mm, the initial delamination length a_0, was produced by using Teflon has a distance 47 mm from the line of load application to the crack tip. The DCB specimens were tested in a tensile testing machine where a tensile load was applied to the specimens through hinges (see **Figure 5**). Immediately before testing, a thin layer of black paint and marked in 1 mm increments were mounted on the specimen to observe crack propagation, starting from the tip of the insert to a length of 47 mm as shown in **Figure 6**. Tests were carried out in a AG-50kNG Shimadzu universal testing machine at 2 mm/min crosshead speed. Following the procedures of ISO 15024:2001, a pre-cracking cycle was performed and given in **Figures 7–10** for different orientation and width of specimen. Initiation values from the insert were then recorded. Further measurements were made in an additional loading cycle, where the crack was allowed to propagate. During the DCB tests, load, displacement, and temperature

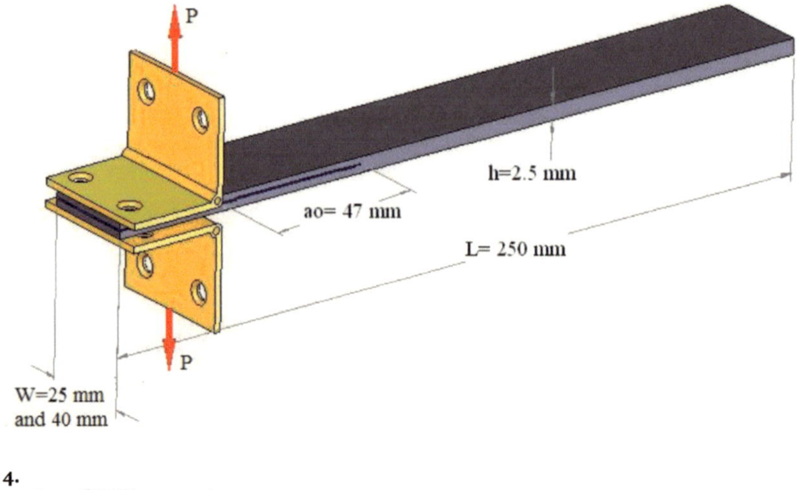

Figure 4.
The dimensions of DCB test specimen.

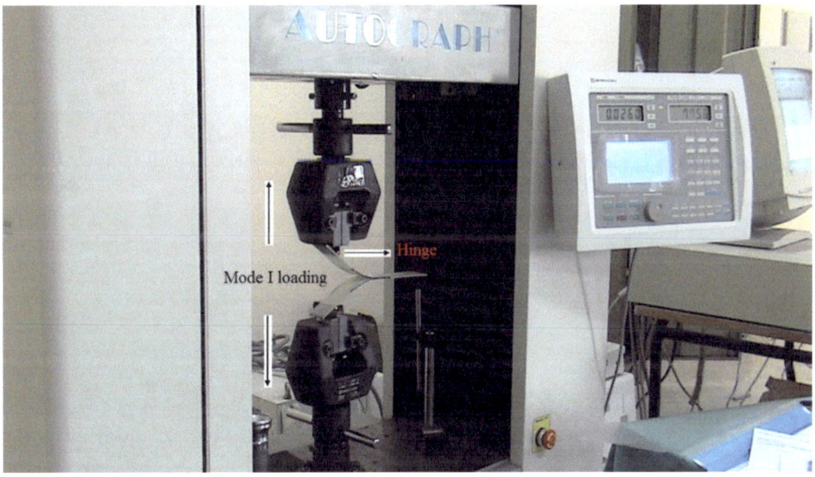

Figure 5.
Experimental setup for DCB specimen and fracture in DCB test.

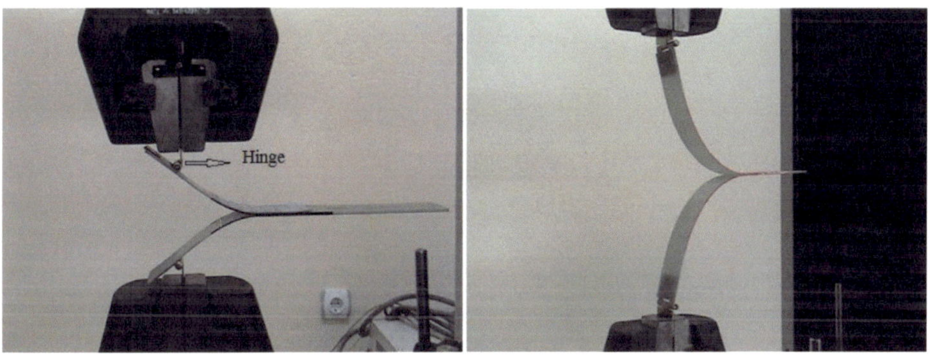

Figure 6.
Overview of loading device and test setup of DCB specimen and delamination a) initially b) progress.

measurements were recorded and video images of the delamination growth were also recorded.

The specimen was subjected to displacement controlled loading and usually experiences stable delamination growth allowing several values of interlaminar fracture toughness to be determined along the specimen's length. As the delamination grows, fiber bridging usually occurs increasing the energy required to propagate further delamination ISO 15025:2001. Interlaminar fracture mechanics was used for characterizing the onset and growth of delaminations and it's calculation is based on experimental CCM as described by ASTM D5528–01. The compliance method is an effective method for determining the fracture characteristics of brittle materials. Fracture toughness is related to the amount of energy required to create fracture surfaces. There are several ways in which initiation and propagation values of G_{IC} can be derived from the recorded load–displacement data. Initiation of delamination is determined by deviation from linearity. G_{IC} can be calculated using the load and displacement at the point of non-linearity of the load–displacement curve. The load–displacement data were then recorded. The unloading curve was also registered, as in case of significant permanent deformations and/or non-linearity. The camera was positioned at a distance 1000 mm away from the specimen surface. Images are captured during the test using two CCD-cameras to get CMOD.

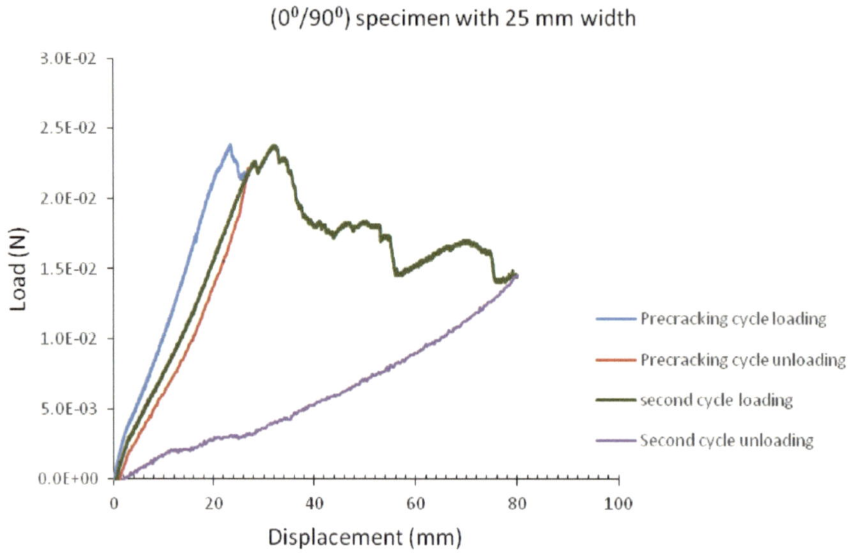

Figure 7.
Load–displacement characteristics for sample 4, woven $[0°/90°]_{16}$ *with 25 mm width.*

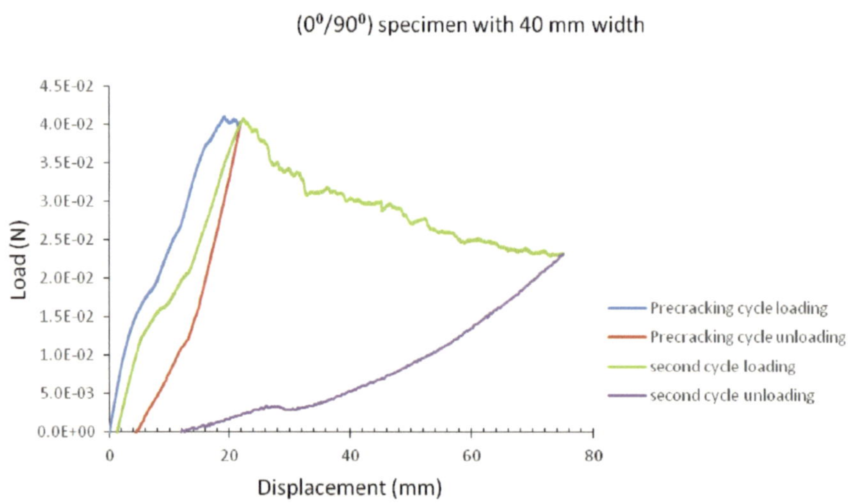

Figure 8.
Load–displacement characteristics for sample 4, woven $[0°/90°]_{16}$ *with 40 mm width.*

In DCB tests, there are four test groups including: *Type 1–1*, *Type 1–2*, *Type 2–1*, *Type 2–2* and they are given in **Table 2**. In *Type X-X* notation, first one denotes direction of woven and second one denotes the width of the DCB specimen. Therefore, *Type 1–1* represents the woven $[0°/90°]_{16}$ with 25 mm width specimen, *Type 1–2* represents the woven $[0°/90°]_{16}$ with 40 mm width specimen, *Type 2–1* represents the woven $[\pm 45°]_{16}$ with 25 mm width specimen, *Type 2–2* represents the woven $[\pm 45°]_{16}$ with 40 mm width specimen. The tests were performed at constant room conditions of 22°C and humidity ratio was 50% in respect of the conditions were needed for standards.

To observe the (delamination) crack growth and to identify crack extension Δa in mm during the loading, the gripped hinges were pulled apart with a crosshead speed of 2 mm/min in displacement control until satisfactory crack

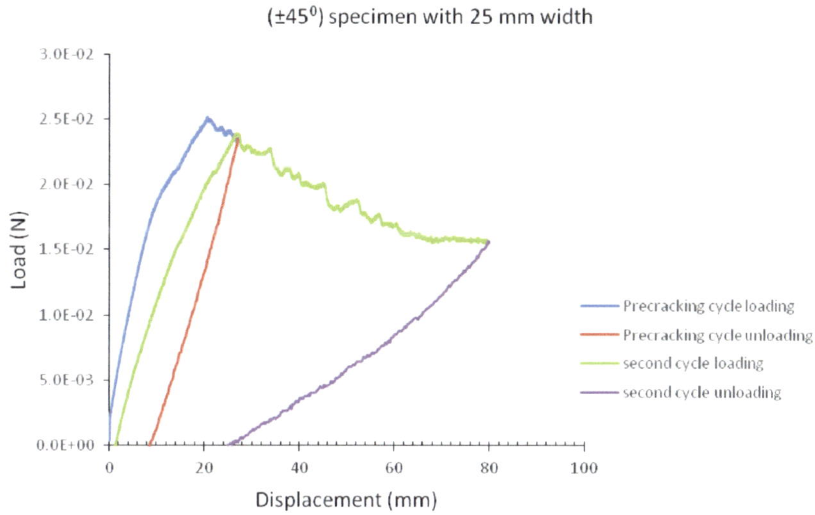

Figure 9.
Load–displacement characteristics for sample 3, woven $[\pm 45°]_{16}$ with 25 mm width.

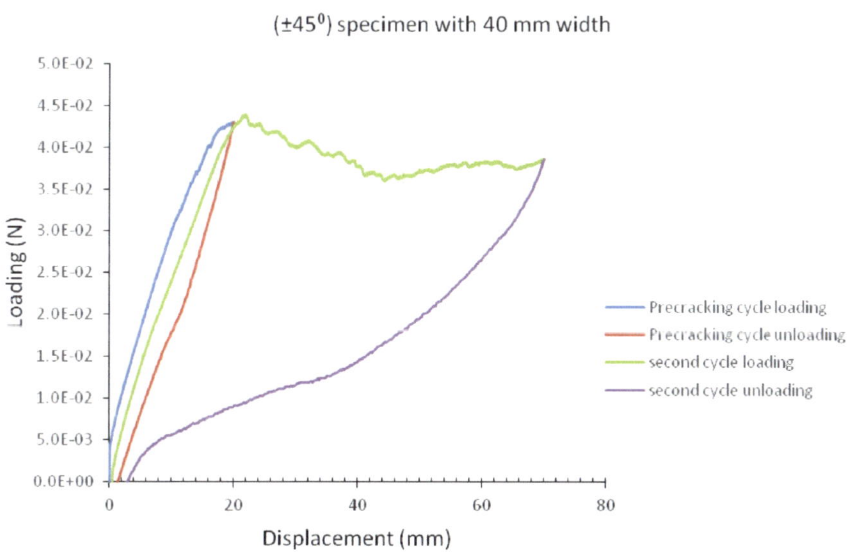

Figure 10.
Load–displacement characteristics for sample 4, woven $[\pm 45°]_{16}$ with 40 mm width.

growth was occurred in the specimen. During the test, the characteristic load–displacement curves were obtained for all test groups and the mean characteristic behavior for each group has been given in **Figures 7–10**. One issue with this test is the slope of the compliance changes as the crack propagates along the specimen. To get the value n, the ratio between the slope of the log c, which is the logarithm of compliance values that is obtained for the system, and log a is the logarithm of the crack length is obtained. Both testing and post processing were conducted similarly for each specimen. First the loading has been obtained with a cross-head speed of 2 mm/min in displacement control up to pre-cracking occurs. Then specimen was unloaded at a constant cross head rate of up to 25 mm/min, the position of the tip of the pre-crack on both edges of the specimen was marked and the procedure was repeated to extent the crack further. Subsequent

DCB specimen type	Fiber direction	Specimen width *b* (in mm)	Total number of *DCB* specimens used in experiment	Obtained average max load (*N*)	Obtained average G_{IC} (J/mm^2)
Type1–1	woven $[0°/90°]_{16}$	25	5	26.48	0.586
Type1–2	woven $[0°/90°]_{16}$	40	5	40.78	0.563
Type2–1	woven $[\pm 45°]_{16}$	25	5	23.18	0.544
Type2–2	woven $[\pm 45°]_{16}$	40	5	33.72	0.530

Table 2.
Fiber directions and width of DCB test specimens used in experiments and fracture energies in pure mode I obtained by DCB under consideration of ASTM standard D5528–01.

measurements were made in an additional loading cycle, where the delamination was allowed to propagate. The unloading curve was also registered, as in case of significant permanent deformations and/or non-linearity. Data reduction was performed according to CCM.

3. Data reduction to obtain interlaminar fracture toughness and energy release rate

There are many analysis methods available for analyzing DCB data. The compliance is defined as the opening displacement measured at the load application points divided by the applied load. Many of the compliance methods are based upon the expression (assuming a linear load-deflection relation): the interlaminar fracture can be calculated by compliance or compliance calibration method, which assumes linear elastic behavior such as [25].

$$G_{IC} = \frac{P_c^2}{2B}\frac{dc(a)}{da} \tag{1}$$

where; P_c is the critical load, a is crack length, B is the specimen width, and $c = \delta / P$ shows the compliance and δ is the CMOD. The compliance values were used to fit versus curve, leading to the critical energy release rate G_{IC}, which is determined by three methods: MBT (Modified Beam Theory), CCM, MCCM (modified compliance calibration method) differed by not more than 3.1%, none of the them were superior to the others [26]. Therefore, G_{IC} for linear elastic material behavior CCM as;

$$G_{IC} = \frac{nP_c\delta}{2Ba} \tag{2}$$

where; P_c is the critical load, a is crack length, B is the specimen width, δ is load point displacement in mm. n is the ratio between the slope of the log δ_i / P_i, which is the compliance of the system and the index i represents the total number of samples and changes from 1 to 5, and P and δ represent simultaneously measured values of load and displacement, respectively during the test, and log a where a represents the crack length. Using a least squares approximation, the slope of log c versus log a yields the value of n for different types of specimen. The mean compliance calibration graphs for different test groups are obtained and given in **Figures 11–14**. According to Castiglione's principle using the relation between the displacement and strain energy (U) as follows.

$$c = \frac{1}{P}\frac{dU}{dP} \tag{3}$$

The CCM was applied to measure the crack growth in each test. Mode I interlaminar fracture toughness values were calculated by means of the following fracture toughness equation:

$$K_{IC} = \sqrt{EG_{IC}} \tag{4}$$

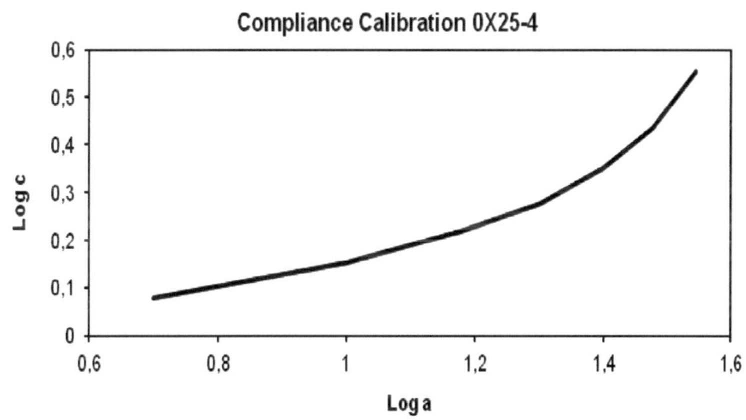

Figure 11.
Curve for slope of log c vs. log a, 'n' for sample 4 woven [0°/90°]$_{16}$ with 25 mm width.

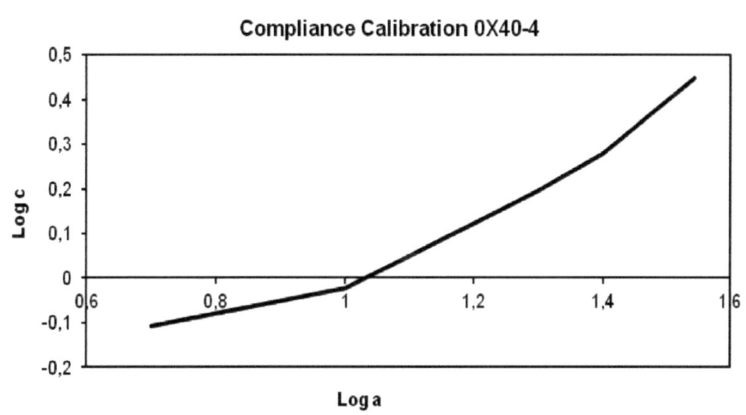

Figure 12.
Curve for slope of log c vs. log a, 'n' for sample 4 woven [0°/90°]$_{16}$ with 40 mm width.

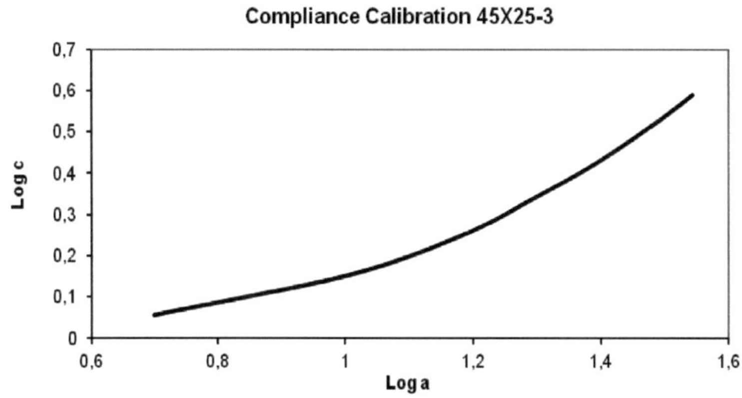

Figure 13.
Curve for slope of log c vs. log a, 'n' for sample 3 woven $[\pm 45°]_{16}$ *with 25 mm width.*

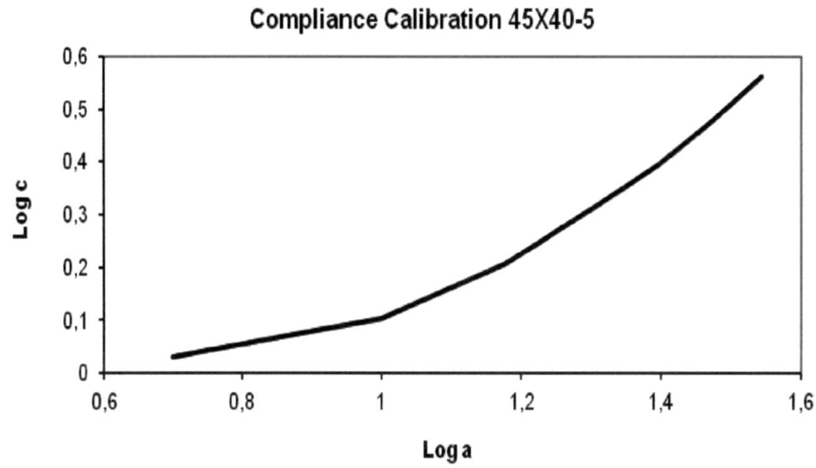

Figure 14.
Curve for slope of log c vs. log a, 'n' for sample 5 woven $[\pm 45°]_{16}$ *with 40 mm width.*

The fracture toughness can be obtained as the function of crack length. In the case of a plane strain condition, the relationship between G_{IC} and K_{IC} is given as follows:

$$G_{IC} = \frac{K_{IC}^2}{E}\left(1-\nu^2\right) \tag{5}$$

where E is the modulus of elasticity, ν is the Poisson's ratio.

The average G_{IC} values are compared and drawn with error bars for 4 types of samples with different width and are given in **Figure 15a** and **b**. As it is seen from the figure, the influence of width on average G_{IC} for [0°/90°] fiber orientation specimens, and [±45°] fiber orientation specimens for different width is small or in other words, as the increasing width of specimen there is a little decrease in average SERR value.

For materials with high interlaminar fracture toughness, it may be necessary to increase the number of plies, that is, increase the laminate thickness or decrease the

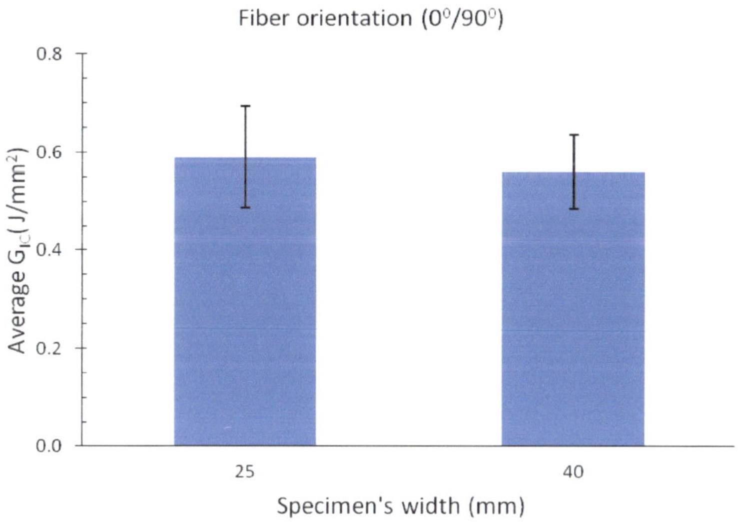

Figure 15.
Influence of width on average G_{IC} for a) [0°/90°] fiber orientation specimens, b) [±45°] fiber orientation specimens. Error bars are ± 1 standard deviations.

delamination length. Thus, for most studies the results indicate a trend of increasing propagation G_{IC} values with increasing DCB thickness [27].

4. Numerical analysis

In the numerical study, the entire specimen was modeled using solid layered 46 element in ANSYS 12.1. Specimens with 25 mm width have 25000 mesh elements

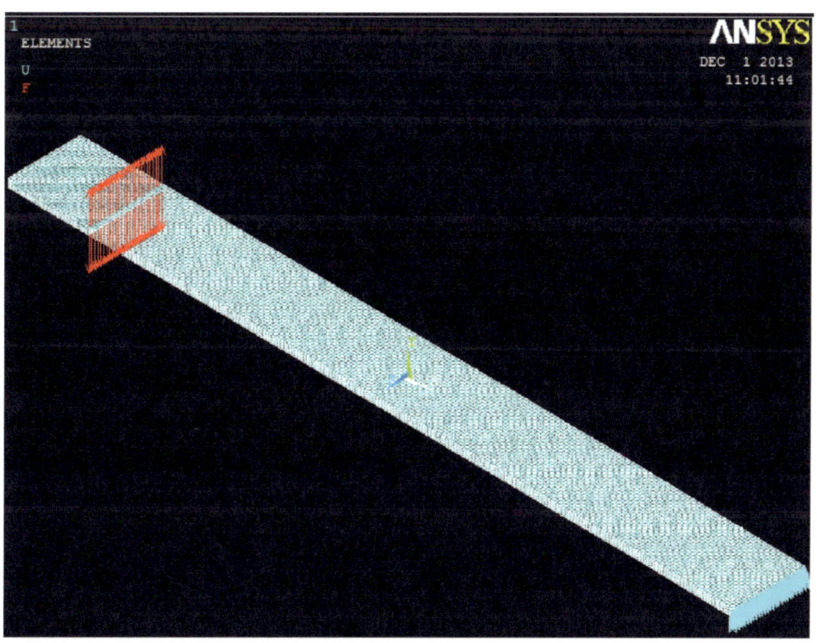

Figure 16.
Detailed view of FEM model of DCB test specimen used under loading.

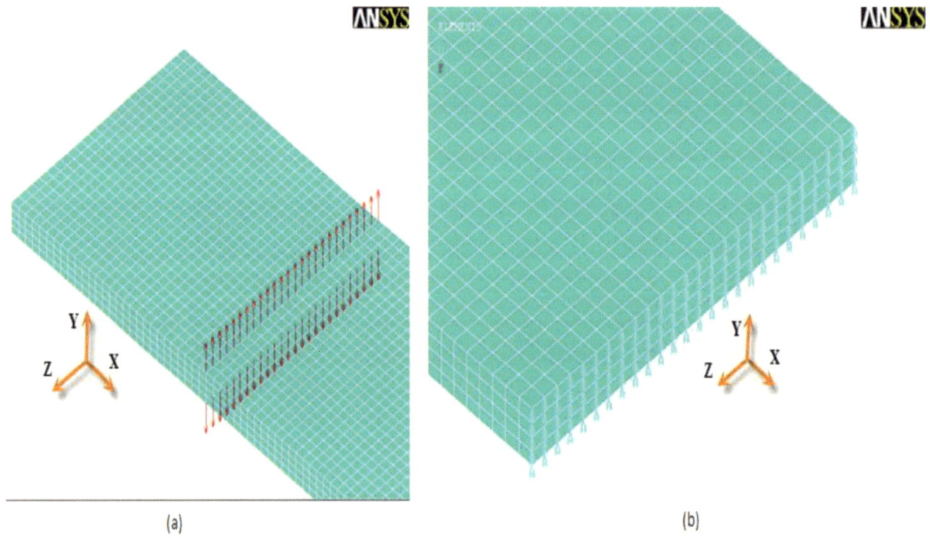

Figure 17.
Forces and boundary conditions of FEM: a) nodal forces, b) boundary conditions.

and specimens with 40 mm width have 40000 mesh elements. A linear elastic finite-element analysis was performed under a plane strain condition. The loads corresponding to the crack initiation in the DCB tests were used to the finite element solution (see in **Figure 16**).

As a part of study, the variation of versus different width length has been investigated. The load was applied as a nodal force at the upper and bottom layer of the composite as shown in **Figure 17(a)** and all nodes along the right side of the composite were restrained in y direction as shown in **Figure 17(b)**. The mechanical properties, which were obtained experimentally, of woven-fabric-reinforced glass/epoxy composite materials are given in **Table 1** and were used during the analysis. The numerical analysis results of the fracture toughness that was obtained by FEM are given in **Table 3**. It is seen that, the obtained numerical values are good comparability with the corresponding experimental results for all experiment groups. As a result of numerical analyses, it can be seen that the fracture toughness increases slightly relative to the amount of decreasing width as mentioned in ASTM D 5528–01 standards, it means that the change of width of DCB specimen is not a great effective parameter for interlaminar fracture toughness values.

The results of experiment tests and numerical analysis showed good agreement demonstrating the effectiveness of the proposed experiment and numerical methods.

DCB specimen type	Average G_{IC} (J/mm^2)	K_{IC} ($MPa\sqrt{m}$) experimental results	K_{IC} ($MPa\sqrt{m}$) ANSYS results
0° × 25 (woven)	0.586	4.076	4.170
0° × 40 (woven)	0.563	3.984	4.110
45° × 25 (45° woven)	0.544	2.600	2.530
45° × 40 (45° woven)	0.530	2.560	2.480

Table 3.
Interlaminar fracture toughness of the woven and woven 45 with different width.

5. Results and discussion

Delamination in composites can occur due to tensile stress Mode I. In the present study, AG-50kNG Shimadzu universal testing machine (see in **Figure 5**) has been used to record the load-deflection curves for calculating fracture toughness for DCB specimen and **Figure 6** shows the initial and the progress of fracture in the machine. Specimens have been designed by referring ASTM standards, DIN EN ISO 75-1, 75–3. The woven $[0°/90°]_{16}$ and woven $[\pm45°]_{16}$ specimens made up of woven-fabric-reinforced glass/epoxy composite materials were studied and the mechanical properties and interlaminar fracture toughness K_{IC} value were obtained experimentally by using DCB and numerically by using FEM. Four types of tests were performed. Each of them includes five tests corresponding to combinations of different stacking sequence of woven and width. For the DCB specimens, the delamination was generated during manufacturing by placing teflon between the mid-layer therefore an initial delamination or crack was formed. In this experiment, the output data have been recorded and graphs have been plotted as required.

The mean values for every individual test groups can be shown in **Figures 11–14**. The SERR, G_{IC} can be derived from the recorded load–displacement data by using compliance calibration method and Eq. (2) to determine Mode I interlaminar fracture toughness K_{IC} given in **Table 3**. The average G_{IC} value of (0°/90°) fiber orientation specimens, for different width with error bars are ±1 standard deviations and the influence of width on average G_{IC} for (±45°) fiber orientation are shown with error bars are ±1 standard deviations in **Figure 15a** and **b**, respectively. It has been obtained that woven $[0°/90°]_{16}$ specimens have higher fracture toughness values than, woven $[\pm45°]_{16}$ specimens to show the effect of different lay-up,.

Figures 7–10 show a typical load–displacement curves for woven $[0°/90°]_{16}$ and woven $[\pm45°]_{16}$ specimens, respectively. According to ASTM 5528–1 the loading process consists of two steps. In step1, the loading was continued with crosshead speed 2 mm/min up to the displacement of about 28 mm and until the first crack propagation Δa 2–3 mm occurs. Then the crosshead was returned to zero point with 25 mm/min constant displacement speed to form a natural pre-crack. In the step 2, second loading was continued with crosshead speed 2 mm/min up to the displacement of about 80 mm, and after then the crosshead was returned to zero point with 25 mm/min constant displacement speed. It is seen that, some degree of non-linearity and small permanent deformations were visible in the unloading curves of specimens. The non-linearity can be partly attributed to the specimens undergoing relatively large displacements towards the end of the tests.

As it is seen from the **Table 3**, the fracture toughness values decrease 2.20% in percentage for the woven $[0°/90°]_{16}$ specimens 25 mm and 40 mm in width, while the specimen width increases. Similarly, for woven $[\pm45°]_{16}$, the fracture toughness values decrease 1.27% in percentage, while the specimen width increases. In other words, the energy release rate increases slightly with decreasing width length for both type of woven specimens. Also it is seen that, the obtained fracture toughness value of woven $[0°/90°]_{16}$ specimen with 25 mm width is greater 3.93% than woven $[\pm45°]_{16}$ specimen. Similarly, it has been found that the fracture toughness value of woven $[0°/90°]_{16}$ specimen 40 mm in width is greater 2.96% than woven $[\pm45°]_{16}$ specimens 40 mm in width.

The obtained average maximum load and the critical SERR, G_{IC} achieved from Eqs. (1) and (2) for plane strain case, have been given in **Table 2**. In the finite element analysis, the stress intensity factor K_{IC} was obtained for DCB specimens numerically. The comparison of experimental and numerical results of K_{IC} values are given in **Table 3**. Finite element analysis was conducted to validate the closed form solution. Results show a good agreement between analytical solutions, numerical simulation shown as in **Table 3**. As it is seen from the table, the differences between numerical and analytical solutions are approximately 2.31–2.69% percentage for 25 mm width specimen and 3.13–3.16% percentage for 40 mm width specimen. It can be said that, when the width range increases the differences between the numerical and analytical solutions may change small amount because of meshing type at delamination through thickness.

6. Conclusion

The problems of interlaminar performance are discussed along with the technique used to measure them and the fracture mechanics principles applied to improve them. The DCB test is a well established test used to measure interlaminar fracture toughness in reinforced composite materials. In this study, interlaminar fracture toughness values are determined by DCB test method Mode I for woven $[0°/90°]_{16}$ and woven $[\pm45°]_{16}$ glass/epoxy composite materials. For investigation of woven angles and specimens width's effect on fracture toughness have been compared. It has been obtained that woven $[0°/90°]_{16}$ specimens have higher fracture toughness values than, woven $[\pm45°]_{16}$ specimens. Some decreases has been eventuated in fracture toughness values by an increase on width. Also, the energy release rate increases with decreasing width length. Therefore, width effect for DCB specimens on fracture toughness value can be neglected. It is also seen that here its no special effect of width on SERR for DCB test specimens. The agreement between the calculated results and the experimental data shows that the finite element analysis method is reliable.

The thickness of material is an important parameter between the state of plane stress and plane strain, and the fracture toughness value can be obtained. For materials with high interlaminar fracture toughness, it may be necessary to increase the number of plies, that is, increase the laminate thickness or decrease the delamination length.

Author details

Evren Meltem Toygar[1*] and Ahmet Gulakman[2]

1 Department of Industrial Engineering, Dokuz Eylül University, Izmir, Turkey

2 Graduate School of Nature and Applied Sciences, Dokuz Eylül University, Izmir, Turkey

*Address all correspondence to: evren.toygar@deu.edu.tr

References

[1] Aliyu AA, Daniel IM. Effects of strain rate on delamination fracture toughness of graphite epoxy. American Society for Testing and Materials. 1985. p. 336-348

[2] Davidson BD, Kruger R, Konig. Effect of stacking sequence on energy release rate distributions in multidirectional DCB and ENF specimens. Engineering Fracture Mechanics. 1996;55(4):557-569.

[3] Sun CT, Zheng S. Delamination characteristics of double cantilever beam and end-notched flexure composite specimens. Composites Science and Technology. 1996;56:451-459.

[4] Toygar M.E., Toparli M. and Uyulgan B., An investigation of fracture toughness of carbon/epoxy composites. Journal of Reinforced Plastics and Composites. 2006;25(18): 1887-1895.

[5] De Morais AB, De Moura MF, Marques AT, De Castro PT. Mode I interlaminer fracture of carbon/epoxy cross–ply composites. Composites Science and Technology. 2002;62:679-686.

[6] Hyer MW. Stress Analysis of Fiber Reinforced Composite Materials. Boston, Massachusetts: McGraw-Hill, 1998.

[7] Broek D. Elemantary Engineering Fracture Mechanics. Boston, Massachusetts: Kluwer Academeic Publishers, 1996.

[8] Jones RM. Mechanics of Composite Materials. Leviitown, PA: Taylor and Francis, 1999.

[9] O'Brien TK. Fracture mechanics of composite delamination. Composites: ASM International. 2001;21:241-245.

[10] O'Brien TK. Characterization of Delamination Onset and Growth in a Composite Laminate, Composite Materials: Testing and Design. Philadelphia: ASTM STP 775, American Society for Testing and Materials. 1982. p. 140-167.

[11] Tay TE. Characterization and analysis of delamination fracture in composites – An overview of developments from 1990 to 2001. Applied Mechanics Reviews. 2003;56:1-32.

[12] ASTM Standard D 5528-01. Standard test method for mod 1 interlaminar fracture toughness of unidirectional fiber-reinforced polymer matrix composites. Philadelphia: Annual Book of ASTM Standards. 1984;3:1.

[13] BS ISO 15024. Fiber-reinforced plastic composites—Determination of mode-I interlaminar fracture toughness, G_{IC}, for unidirectionally reinforced materials. British: Standard, United Kingdom.

[14] M.N. Durakbasa, P.H. Osanna, P. Demircioglu, The factors affecting surface roughness measurements of the machined flat and spherical surface structures – The geometry and the precision of the surface, Measurement, 2011;44(10), 1986-1999.

[15] DIN EN ISO 75-1, Kunststoffe. Bestimmung der Wärmeformbeständigkeitstemperatur. Teil 1: Allgemeines Prüfverfahren (ISO 751:2004);Deutsche Fassung EN ISO 751:2004, September 2004.

[16] DIN EN ISO 75-3, Kunststoffe. Bestimmung der Wärmeformbeständigkeitstemperatur. Teil 3: Hochbeständige härtbare Schichtstoffe und langfaserverstärkte Kunststoffe (ISO 753:2004); Deutsche

Fassung EN ISO 753:2004, September 2004.

[17] J.H. Chen et al. Effect of fibre content on the interlaminar fracture toughness of unidirectional glass-fibre/ polyamide composite. Composites: Part A 1999;30: 747-755.

[18] Martin RH, Murri GB. Characterization of mode I and mode II delamination growth and thresholds in AS4/PEEK composites. Composite materials: Testing and design. Philadelphia: ASTM STP 1059. American Society for Testing and Materials. 1990;9:251-270.

[19] O'Brien TK. Towards a Damage Tolerance Philosophy for Composite Materials and Structures. Composite Materials: Testing and Design. Philadelphia: ASTM STP 1059, American Society for Testing and Materials. 1990. p.7-33.

[20] Russell AJ, Street KN. A Constant G Test for Measuring Mode I Interlaminar Fatigue Crack Growth Rates. In: Proceedings of Composite Materials: Testing and Design (Eighth Conference). Philadelphia: ASTM STP 972, American Society for Testing and Materials. 1988. p.259-277.

[21] Whitney JM, Browning CE, Hoogsteden W. A double cantilever beam test for characterizing mode I delamination of composite materials. Reinforced Plastics and Composites 1982; 1: 297-313.

[22] ASTM Standard E6-81. Standard Definitions of Terms Relating to Methods of Mechanical Testing. Annual Book of ASTM Standards. Philadelphia: ASTM, 1981. p.187-196.

[23] Zhang Z. Research and Development Facilities in Advanced Polymer and Composites (APC) Research Group. Portsmouth, UK, 2010. http://www.port.ac.uk/research/composites/filetodownload,106518,en.pdf (Accessed 06 April 2013).

[24] Reder C., Loidl D., Puchegger S., Gitschthaler D., Peterlik H., Kromp K., Khatibi G., Betzwar-Kotas A., Zimprich P., Weiss B., Non-contacting strain measurements of ceramic and carbon single fibres by using the laser-speckle method, Composites: Part A 2003; 34:1029-1033.

[25] Anderson TL. Fracture Mechanic: Fundamental and Applications. CRC Press, 2nd edition, 1995.

[26] M.S. Sham Prasad, C.S. Venkatesha, T. Jayaraju. Experimental methods of determining fracture toughness of fiber reinforced polymer composites under various loading conditions. Journal of Minerals & Materials Characterization & Engineering, 2011;10(13), 1263-1275.

[27] Hojo M., Aoki T. Thickness Effect of Double Cantilever Beam Specimen on Interlaminar Fracture Toughness of AS4/PEEK and T800/Epoxy Laminates in, Composite Materials, Fatique and Fracture, Fourth Volume, ASTM STP 1156, W. W. Stinchcomb and N. E. Ashbaugh (Eds), ASTM, Philadelphia, 2007;281-298.

Chapter 3

A Probabilistic Approach in Fuselage Damage Analysis *via* Boundary Element Method

Gilberto Gomes, Thiago Oliveira and Francisco Evangelista Jr

Abstract

This chapter presents a new alternative approach to the analysis of the fatigue life of aircraft fuselage parts considering the compliance of internal elements to replace the classical model of critical crack size. In this case, from a global–local analysis using the boundary element method (BEM), induced stresses at a macro model, and their effects on micro models are evaluated. The BEM enables efficient simulations of the propagation of initial defects to assess the damage tolerance. For this purpose, computational techniques were developed that allowed evaluating these models, through a probabilistic treatment to assess damage tolerance and fatigue life. Finally, this technique is shown as an alternative to ensure the integrity and proper operation of fuselage panels avoiding reaching a Limit State during its projected lifespan.

Keywords: Fatigue, aircraft fuselage, compliance, global–local analysis, Boundary Element Method, damage tolerance

1. Introduction

Some documented studies to interpret the cause of aircraft accidents [1, 2]. One of the classical cases is the accidents with the Comet aircraft and Boeing 737–200 [2]. Regarding the Comet accident, the reports concluded fatigue failure as the main reason for the disintegration of the pressurized cabin. This study also reported that although designed for the operating conditions, its structure was unable to prevent the crack propagation, particularly unstable cracks that after reaching the critical length would continue to propagate until complete rupture of the structure.

As for the Boeing 737 Aloha Airlines, it stabilized at 7,000 meters altitude as planned, and a loud bang suddenly followed by the disintegration of the roof leaving a gap of six meters in the fuselage in flight, but it was still possible to land the aircraft with a damaged structure as shown in **Figure 1**. In this case, the investigations indicated that fatigue crack initiation in several areas (multiple damage site - MSD) greatly reduced the strength of the structure causing it to collapse.

Designers are always looking for fast and reliable simulation methods that produce accurate results to avoid damage processes and, consequently, the occurrence of accidents. Automation is then seen as a key point to evaluate several scenarios for parametric studies resulting in design optimization [3]. Thus, numerical methods as

Figure 1.
Boeing 737–200 fuselage fatigue failure. NTSB. Aircraft Accident Report NTSB/AAR-89/03, Aloha Airlines, Flight 243, Boeing 737–200, N73711; Near Maui, Hawaii; April 28, 1988.

domain methods (Finite Element Method – FEM, Extended Finite Element Method – XFEM, Generalized Finite Element Method - GFEM), contour methods (Boundary Element Method – BEM, Dual Boundary Element Method – DBEM, Radial Integration Method - RIM) and mesh-free methods appear as an alternative for solving fracture problems. The DBEM shows further advantages simplifying the modeling of the crack area, direct SIF calculation, run times reduced, and accurate simulation of crack growth [4–6]. The need for discretization only of the solid contour allows, using the DBEM, the analysis of thousands of probabilistic and reliability simulations, such as flaws, initial defects, fatigue behavior prediction, multiple local damages, among others [7–11].

The literature analyzes the damage tolerance (number of cycles) based on the crack size. This chapter presents an innovative method for damage tolerance based on probabilistic global–local analysis. In this context, the method can find a relationship between the number of fatigue cycles and the respective compliance in local elements considering the statistical nature of the input parameters.

2. Literature review

2.1 Damage tolerance

Over the years several fatigue design methodologies have developed trying to combine structural safety and low cost in the aircraft manufacturing and operation process. The first methodology was called safe-life. This approach consists of designing and manufacturing a safe aeronautical structure throughout its useful life. For this, one must consider, in the prototype tests, the most extreme situations of fatigue stresses, foreseen during operation. Such methodology results in factors that oversize the structural elements to prevent the possibility of failure. This approach leads to high project costs and is not able to guarantee safety if an unforeseen design failure occurs during its useful life.

Rationally, a new methodology was developed based on the concept of damage tolerance. In this methodology, it is assumed that the structure is capable of withstanding the actions for which it was designed until the detection of a fatigue crack or other defects during its operation. The aircraft is then checked, repaired, and put back into service until the end of its useful life. The concept of damage tolerance began from statistical analysis to control the spread of fatigue cracking and considering inspection intervals to maintain a low probability of complete failure [12]. Later, the damage tolerance has been applied in the use of aluminum alloys for aircraft structural applications [13, 14]. Thus, the fatigue damage analysis considered some aspects of the design, predictions, and experiments associated with tolerance to damage to aircraft structures [15–17]. From this, it was noticed that the load cycles have a direct linear relationship with the logarithm of the crack size and that the largest formed cracks grow in an approximately exponential way, known as the "main crack" methodology [18], from small discontinuities (flaws and microcracks) inherent to the material, as soon as an aircraft enters service [19]. Currently, the concept of damage tolerance is applied to aircraft with composite structures [20–23], in the analysis of multiple cracks [24], and shape optimization design [25, 26]. Studies on tolerance to probabilistic damage are based on manufacturing components [27] and the dispersion of fatigue life from the distribution of initial defects [28]. Other works relate damage tolerance through computational methods, using XFEM [29, 30], BEM [31], and DBEM [32].

2.2 Fatigue

Fatigue is characterized by a cyclic loading process that causes progressive internal structural cumulative damage. In this case, the Paris Law [33], Eq. (2), relates the crack propagation rate (*da/dN*) with the variation of the Stress Intensity Factor (ΔK):

$$\frac{da}{dN} = C.\Delta K^m \tag{1}$$

where a is the crack length, N is the number of load cycles, C and m are material dependent constants. According to classical theory, after a certain number of cycles, the cracks reach a critical length making the structure unstable and causing it to collapse. Thus, developing the Eq. (1), the condition for critical crack size a_c is determined by:

$$N = \frac{1}{C}\int_{a_0}^{a_c} \frac{da}{\Delta K^m} > N_{tot} \tag{2}$$

where N is the number of cycles required to increase the crack of the initial size a_0 up to a critical crack length a_c and N_{tot} is the number of cycles throughout life.

2.3 Boundary element method

The Finite Element Method (FEM) is probably the main technique used in engineering analysis, standing out due to its great versatility, quality of results, and relative ease to implement [34]. On the other hand, the Boundary Elements Method (BEM) has emerged as a complementary alternative to FEM, being indicated particularly in special cases that require better interpretation and data representation in problems with stress concentration or where the domain is infinite or semi-infinite [35].

Thus, the boundary elements technique began to be used in problems of incremental crack extension analysis [36]. The solution of crack problems, in general, cannot be achieved in an analysis of a single region with direct application of the conventional BEM, because the application of the same boundary integral equation at coincident source points on the opposing crack surfaces leads to degeneracy in the resulting system of algebraic equations. Among the techniques applied to work around this problem are the sub-regions, which models the structure in artificial contours connecting the cracks to the boundary in such a way that the domain is divided into sub-regions without cracks [5], as well as the Dual Boundary Elements

Method (DBEM), based on two distinct integral equations on each crack face (displacement and traction equations). Thus, degeneration of the equations system generated by the BEM is no longer present and the need for remeshing vicinity of the crack tip is not required, generating only new rows and columns to the existing matrix [6].

BemCracker2D is a program for elastostatic analysis of 2D problems to performing analyzes using the Boundary Element Method [37–41]. This software performs modeling of the standard BEM and the incremental analysis strategy for problems involving cracks [42]. For the analysis of the fracture mechanics problems, the BemCracker2D calculate elastic stresses using the conventional BEM and performs incremental analyzes of the crack extension through the DBEM.

3. Methodology

Initially, by submitting the macro model to the normal load P and shear load Q, the stress fields on the plate are obtained from the analysis of the continuum mechanics regarding the Eqs. (3), as shown in **Figure 2**.

Thus, analytically, the stress fields in a body in the elastic regime submitted to normal and shear stress, as shown in **Figure 2**, are represented in Eqs. (3):

$$\begin{aligned}
\sigma_x &= -\frac{2z}{\pi}\int_{-a}^{b}\frac{p(s)(x-s)^2}{\left((x-s)^2+y^2\right)^2}ds-\frac{2}{\pi}\int_{-a}^{b}\frac{q(s)(x-s)^3}{\left((x-s)^2+y\right)^2}ds\\
\sigma_y &= -\frac{2z^3}{\pi}\int_{-a}^{b}\frac{p(s)}{\left((x-s)^2+y^2\right)^2}ds-\frac{2y^2}{\pi}\int_{-a}^{b}\frac{q(s)(x-s)}{\left((x-s)^2+y^2\right)^2}ds\\
\tau_{xy} &= -\frac{2z^2}{\pi}\int_{-a}^{b}\frac{p(s)(x-s)}{\left((x-s)^2+y^2\right)^2}ds-\frac{2y}{\pi}\int_{-a}^{b}\frac{q(s)(x-s)^2}{\left((x-s)^2+y^2\right)^2}ds
\end{aligned} \tag{3}$$

With the stress field obtained in the macro model, analyzes are performed in the microelement. The microelement is represented by a square region of unitary side with a load on the right and top edges, and supported on the left and bottom edges, with a central hole and two 45° inclined cracks representing initial defects in the piece, as shown in **Figure 3**. Stresses σ_x, σ_y, and τ from the stress field at an internal point of the macro analysis, applied at the load ratio R = 0.5.

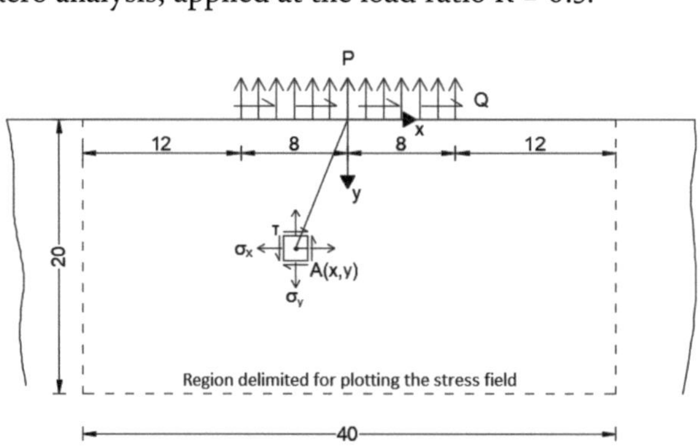

Figure 2.
Global–local analysis: Macro model with loadings and dimensions; and detail of the stress in the microelement (units in cm).

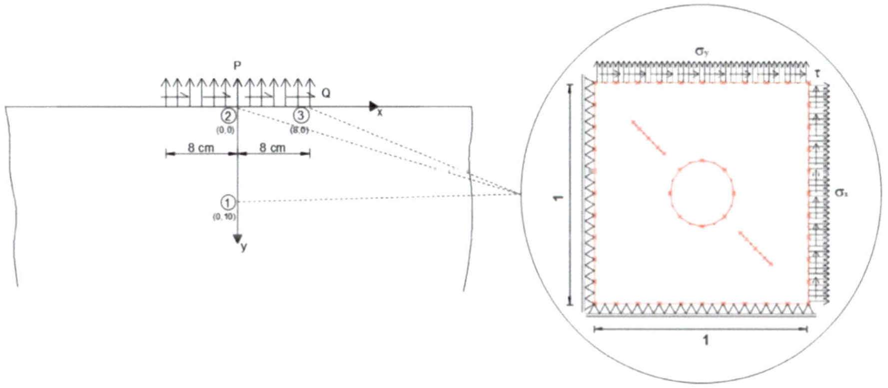

Figure 3.
Macro model and positions for initial defect analysis (microelements) and detail of the microelement for the microanalysis (units in cm).

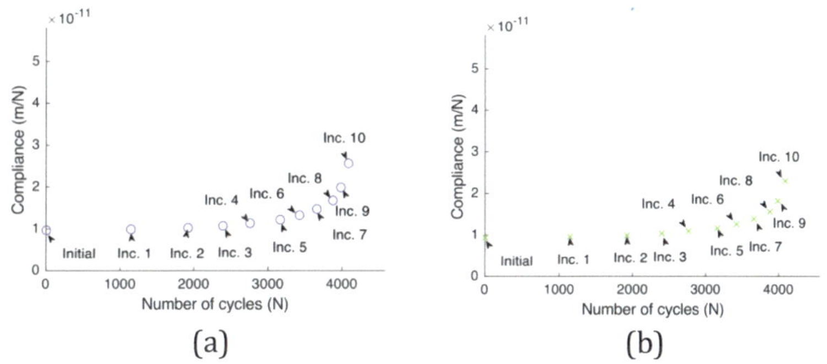

Figure 4.
Points of the number of cycles versus *compliance ratio at each increment. (a) Top edge. (b) Right edge.*

The microanalysis was performed in three different positions of the macro element. Position 1 considers the microelement in the center of the plate with coordinates (0.10); position 2 considers this element at the origin of the system axis (0,0); and position 3, at the limit of the application of the external request (8.0), such positions are illustrated in **Figure 3**.

When performing the analysis on the microelement, BemCracker2D brings, as a result, the number of cycles and the displacement of the boundary mesh for each crack propagation increment. With these results, the compliance is calculated from the average of the displacements of each edge and the respective stress on the considered edge (right or top), thus obtaining the points to form the curve of the number of cycles *versus* compliance, as shown in **Figure 4**.

Each point (from left to right) represents an increase in crack size. Initial compliance is on the abscissa axis. As the cracks spread, the plate loses rigidity, exponentially increasing the compliance *versus* cycles ratio. From these points, a spline curve is fitted as seen in **Figure 5**. With the curves, the number of cycles is obtained, which corresponds to the compliance C (C_x and C_y) two times initial compliance 2C ($2C_x$ and $2C_y$), and three times initial compliance 3C ($3C_x$ and $3C_y$).

The results of the number of cycles were evaluated for the four crack tips of the microelement as shown in **Figure 6**, where C1 T1 means Crack 1 Tip 1; C2 T2, Crack 2 Tip 2, and so on, with the lowest number of cycles being considered to achieve the results for the 2C and 3C.

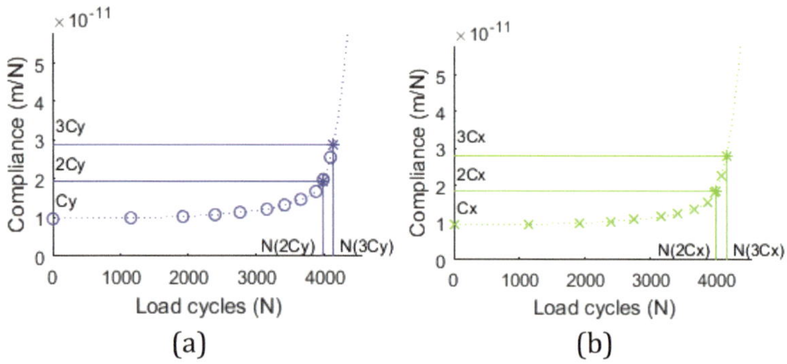

Figure 5.
Number of cycles versus *compliance curves. (a) Upper edge. (b) Right edge.*

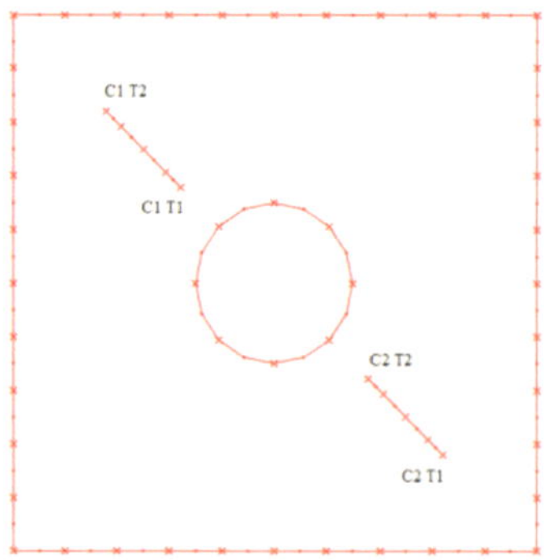

Figure 6.
Crack tips.

Structural aircraft fuselage commonly uses the 2024 aluminum alloy as a base material due to its high ability to withstand damage, good mechanical strength, and corrosion resistance [43, 44]. Thus, the material considered for the fuselage panel was aluminum alloy 2024-T3, yield and limit strengths were 338 MPa and 476 MPa, respectively, Young's modulus and Poisson's coefficient of 74 GPa and 0.33, respectively, and fracture toughness (K_c) of 34 MPa$\sqrt{m}$ [45]. The Paris Law's parameters C = 7.20e-11 and m = 3.52 [46] were considered with the fatigue load ratio being R = 0.5.

The proposed analysis considered the effects on fatigue life by changing the values of the following variables: external loadings P (normal) and Q (shear) the macro analysis; contained in the Paris Law C and m; and initial defects R (hole radius), L1 (upper crack) and L2 (lower crack), from the microanalysis.

The following is an example of the technique, considering the values of the variables represented in **Table 1**:

P and Q are the normal and shear stresses, respectively, C and m are the Paris constants, r the radius of the central hole, L1 and L2 the size of the upper and lower cracks, respectively. For these values, we obtain the stress fields of the macro analysis illustrated in **Figure 7**.

Position 1.

Table 2 shows the resulting stresses considering the microelement at position 1 of **Figure 3**.

Figure 8(a) shows the 10 crack propagation increments, while **Figure 8(b)** shows the deformed microelement after all the increments, highlighting two details:

P (MPa)	360.47
Q (MPa)	92.78
C	7.20e-11
m	3.52
r (cm)	0.093
L1 (cm)	0.086
L2 (cm)	0.093

Table 1.
Values of case study 1 variables.

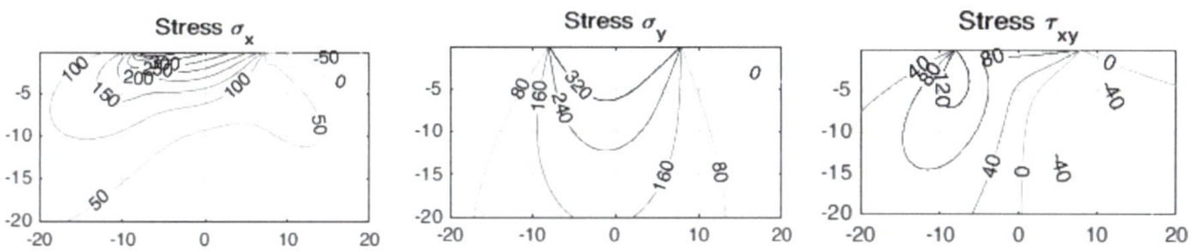

(a) σx (b) σy (c) τ

Figure 7.
Calculated stress field.

σ_x	42.90 MPa
σ_y	266.78 MPa
τ	11.04 MPa

Table 2.
Stress results at position 1.

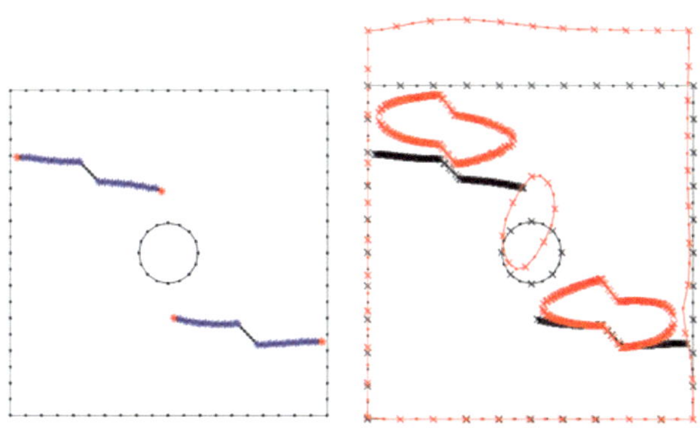

(a) Crack propagation (b) Deformed mesh

Figure 8.
Analysis for the microelement in position 1.

1. Since normal σ_y stress has a magnitude higher than the σ_x (**Table 2**), strains were much higher in y to x, deformation still occurs contrary to the direction of applied stress (tensile and σ_x deformation in the negative direction of the x-axis) due to Poisson effect. This implies disregarding this negative compliance for the analysis. Therefore, **Figure 9** considers only the analysis for the upper border.

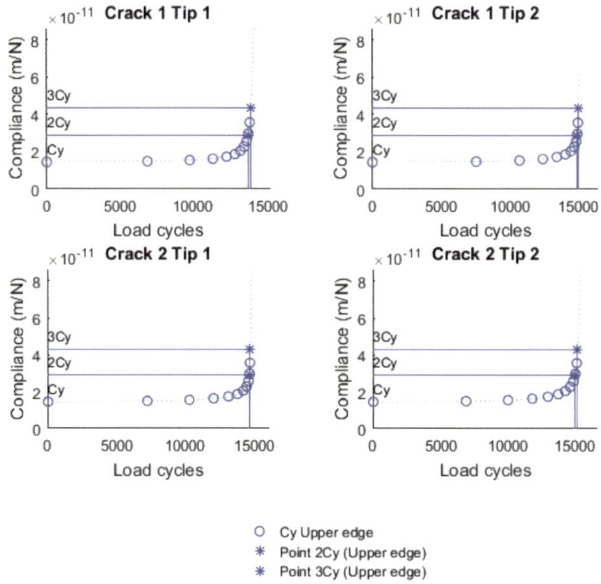

Figure 9.
Number of cycles x compliance for the micro element in position 1.

σ_x	349.00 MPa
σ_y	360.47 MPa
τ	89.83 MPa

Table 3.
Stress results at position 2.

2. The analysis was able to detect the loss of local stiffness near the crack zones. At the upper edge, the microelement near the upper crack had a much greater displacement than at the other points; the same occurring on the lateral edge close to the lower crack.

The smallest number of cycles to reach 2Cy and 3Cy, we have N(2C) = 1.3635e+04 and N(3C) = 1.3811e+04, respectively.

Position 2.

Table 3 shows the resulting stress field considering the microelement at position 2 of **Figure 3**.

Figure 10(a) shows the crack increments. This analysis also resulted in ten propagation increments. **Figure 10(b)** shows the deformed microelement after all the increments. As a result, it can be seen that as the stresses have similar magnitudes (σ_x = 349.00 MPa and σ_y = 360.47 MPa), the deformations in the element have a certain symmetry. Thus, the two edges that have compliance, should be assessed the lowest number of cycles for each crack tip resulting in the first value of 2C and 3C. The result points N(2C) = 1.8601e+03 and N(3C) = 1.8636e+03, as shown in **Figure 11**.

Position 3.

Table 4 shows the resulting stress field considering the microelement at position 3 of **Figure 3**.

Figure 12(a) illustrates the crack increments. This analysis also resulted in 10 propagation increments. Then, **Figure 12(b)** represents the deformed microelement after all the increments. As a result, it is noticed that as σ_x has a magnitude

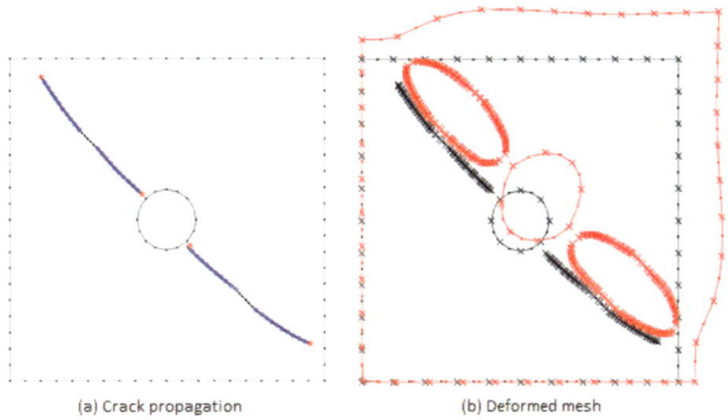

Figure 10.
Analysis for the microelement at position 2.

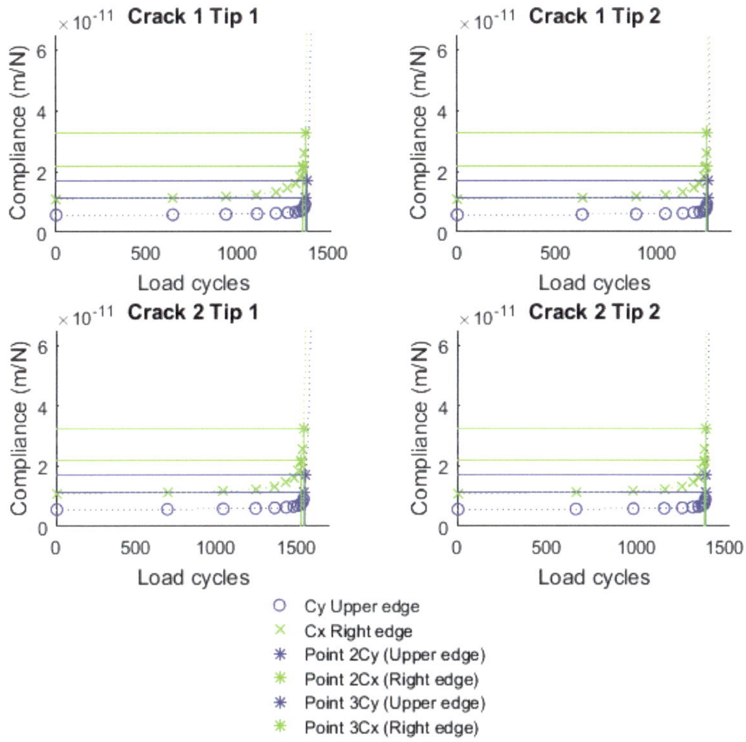

Figure 11.
Number of cycles x compliance for the microelement at position 3.

σ_x	406.67
σ_y	209.76
τ	160.38

Table 4.
Stress field (MPa) at position 3.

much greater than σ_y, the element has more prominent deformation on the right edge. In this analysis, the two edges also show compliance, and the lowest number of cycles for each crack tip that results in the first value of 2Cx and 3Cx must be evaluated. The result points N(2C) = 1.2467e+03 and N(3C) = 1.2521e+03, as shown in **Figure 13**.

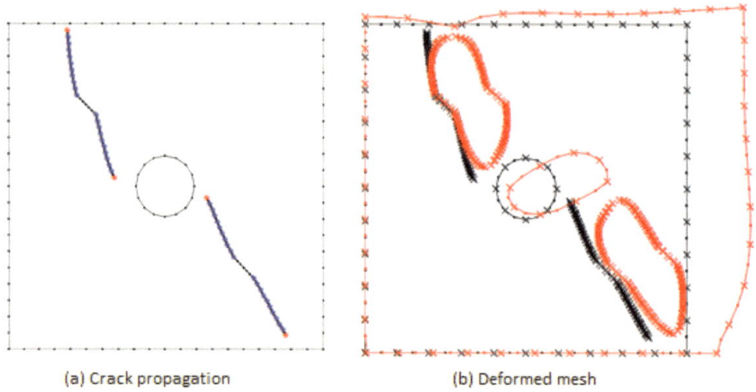

Figure 12.
Analysis for the microelement at position 3.

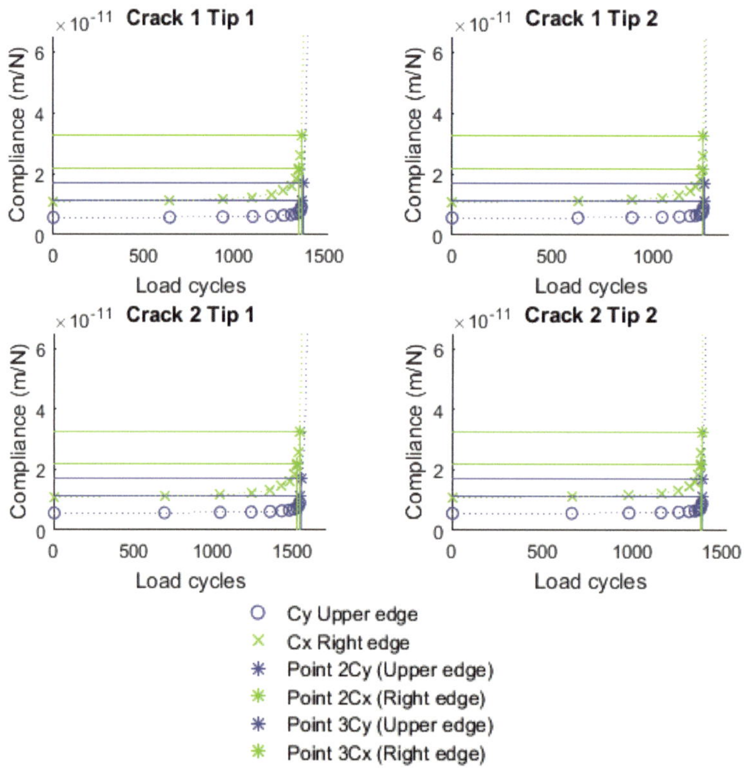

Figure 13.
Number of cycles versus *compliance for the microelement at position 3.*

4. Probabilistic analysis

The values obtained previously refer to a deterministic analysis. A probabilistic analysis using Monte Carlo (MC) sampling performed a thousand simulations using BEM varying the values of P, Q, C, m, R, L1, and L2. The statistical parameters for those variables used for the MC simulation are listed in **Table 5**.

The results are presented from **Figures 14–16. Figure 14** refers to the results for microelement in Position 1. It can be seen that the instability of the microelement occurs for load cycles varying between 10^3 and 10^6. **Figure 15** refers to the results

Variable	Mean	Standard deviation
C	7.0e-11	8.5e-12
m	3.2	0.4
P (MPa)	301.2	44.1
Q (MPa)	99.1	15.1
R (cm)	0.1	1.2e-02
L1 (cm)	0.1	9.8e-03
L2 (cm)	0.1	1.0e-02

Table 5.
Mean and standard deviation values for each variable.

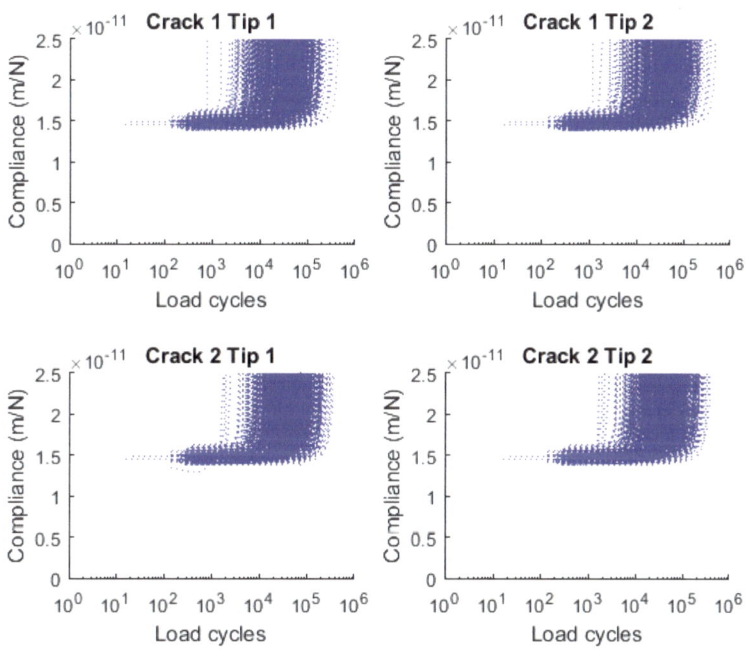

Figure 14.
Monte Carlo simulations of compliance versus *number of cycles in position 1.*

for the microelement in Position 2. It can be seen that the instability of the micro-element occurs for load cycles varying between 10^2 and 10^5. **Figure 16** refers to the results for Position 3. It can be seen that the instability of the microelement occurs for small cycles of load ranging between 10^1 and 10^5.

It can be noted that when the compliance reaches the value of three times the initial compliance (3C), the element already becomes unstable and tends to increase infinitely as can be seen in **Figures 16–18**. **Table 6** shows the minimum number of cycles *N* that leads the microelement to reach 3C, thus occurring instability. The smallest *N*-value is the worst analysis case, which is the most conservative case. For Position 1 the element becomes unstable in 1.113e+03 cycles. For Position 2, instability occurs in 1.12e+02 cycles, in this case, there is a sudden reduction of load cycles regarding the previous case (ten times less than the element in Position 1). In position 3, the instability occurs in 4.7e+01 cycles, being the worst situation in the case of occurrence of initial defects in the panel. **Figure 17** shows the superposition of the maximum and minimum limits of all curves in each position.

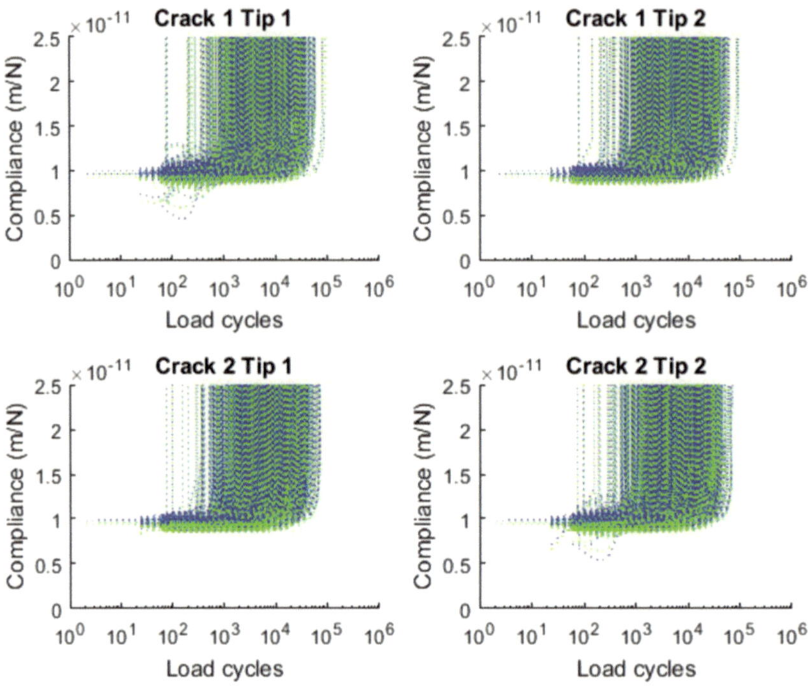

Figure 15.
Monte Carlo simulations of compliance versus *number of cycles in position* 2.

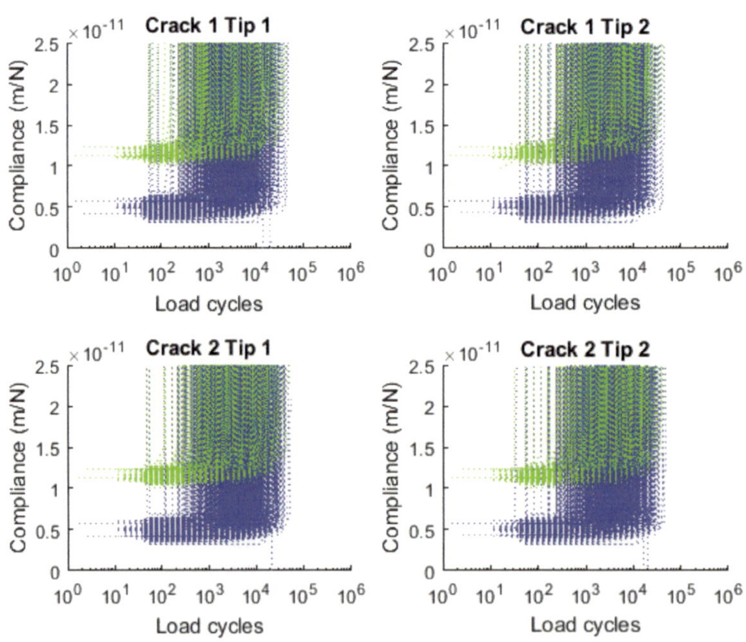

Figure 16.
Monte Carlo simulations of compliance versus *number of cycles in position* 3.

Figure 18 shows the Absolute Frequency (AF) of the incidences of compliance values corresponding to the initial compliance C, 2C, and 3C. **Figure 19(a)–(c)**, for the microelement, show the positions 1, 2, and 3, respectively. Note the increase in the dispersion of the value distribution due to higher standard deviation observed for increasing compliance levels.

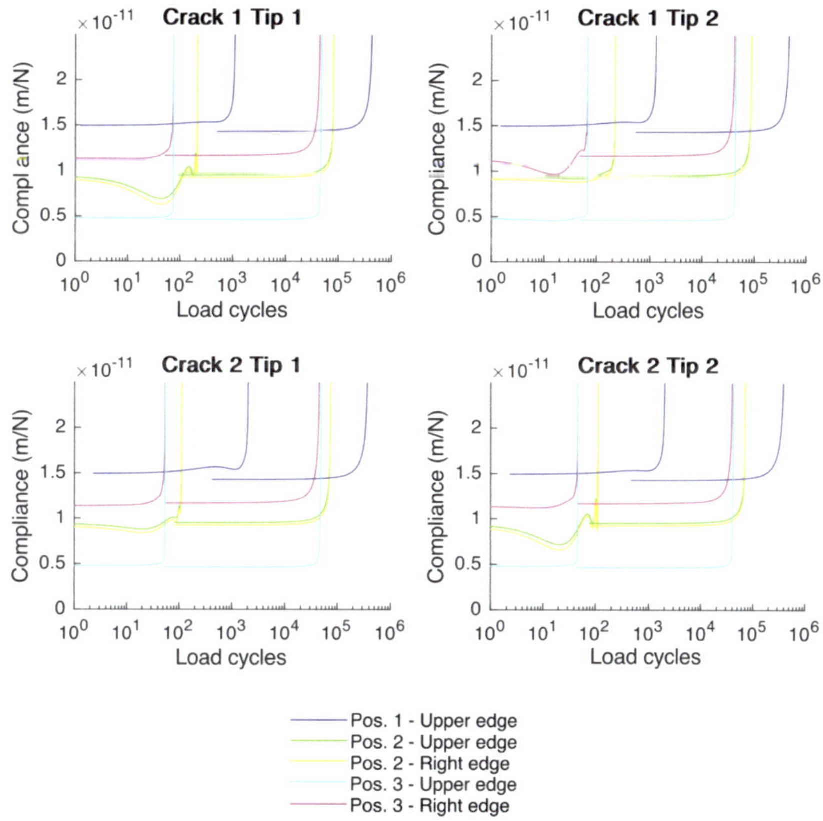

Figure 17.
Compliance versus *number of cycle limits in positions 1, 2, and 3.*

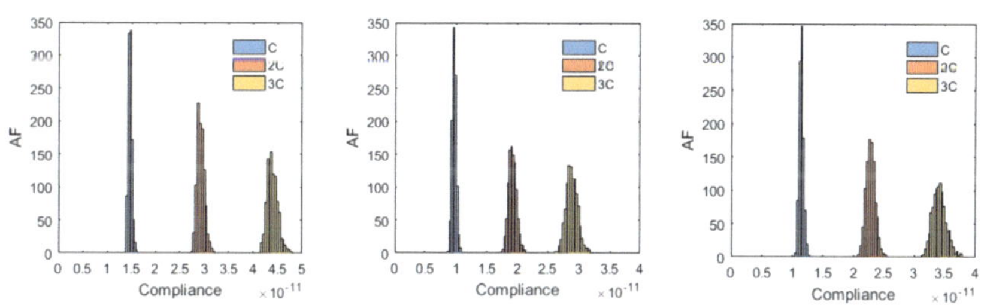

(a) Position 1 (b) Position 2 (c) Position 3

Figure 18.
Absolute frequency of the incidences of compliances values corresponding to the initial compliance C, 2C, and 3C for the three positions.

Position 1	Position 2	Position 3
1.113e+03	1.12e+02	4.7e+01

Table 6.
Minimum number of cycles that lead the microelement to instability.

Figure 19 shows the probability density function $f(N_{3c})$ to the micro element damage tolerance analysis for the N cycles at 3C. It is noticed that the element in position 3 has a marked curvature and consequently a lesser variability of damage tolerance, that is, a selection range of N which is limited to 3C is limited to 0.4e+05. The element in position 2 has a variation of N that reaches 3C up to 0.7e+05.

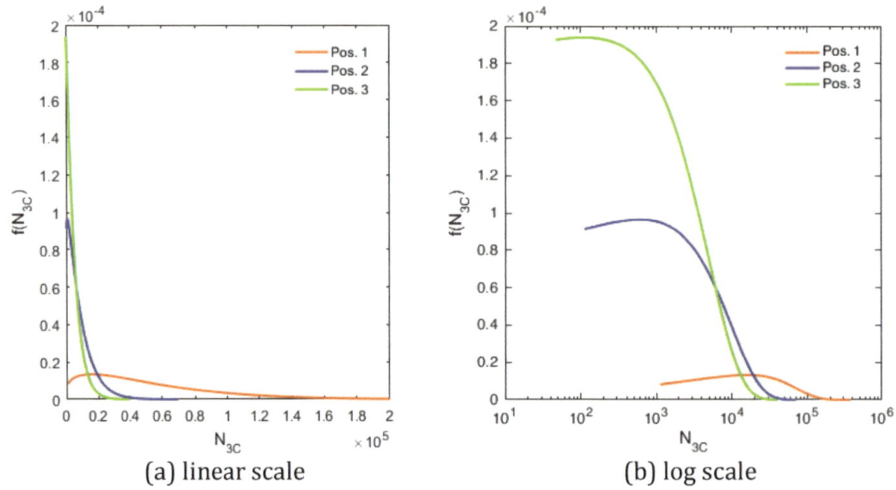

Figure 19.
Estimated probability density functions for fatigue life for 3C criterion.

	Mean	Standard deviation	Coefficient of variation
Position 1	5.1e+04	4.5e+04	0.8823
Position 2	8.5e+03	9.0e+03	1.0544
Position 3	4.7e+03	5.1e+03	1.0894

Table 7.
Mean values and standard deviation of number of cycles to 3C.

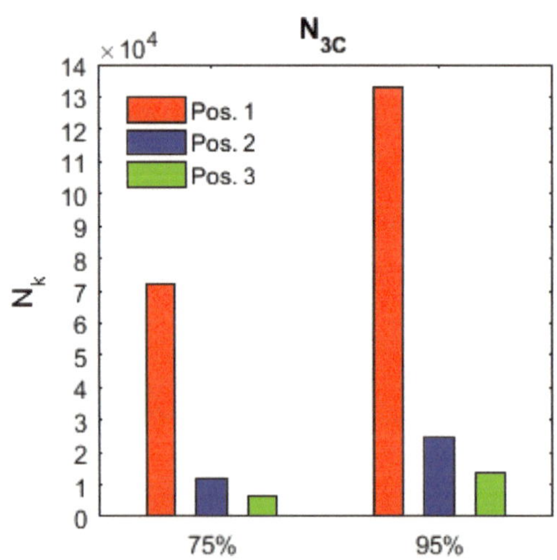

Figure 20.
Characteristic value for fatigue life (N_k) considering for 3C criterion.

Position 1 presents a greater dispersion, which means a greater variation of the values of N_{3c}. This is confirmed in **Table 7** by the highest coefficient of variation.

Considering the determination of the characteristic fatigue life (N_k), related to an α% probability that an N value would not be superior to N_k, **Figure 20** shows those characteristic values for that for α of 75% and 95%. Note the significantly higher values for position 1 rather than the other two positions for both α.

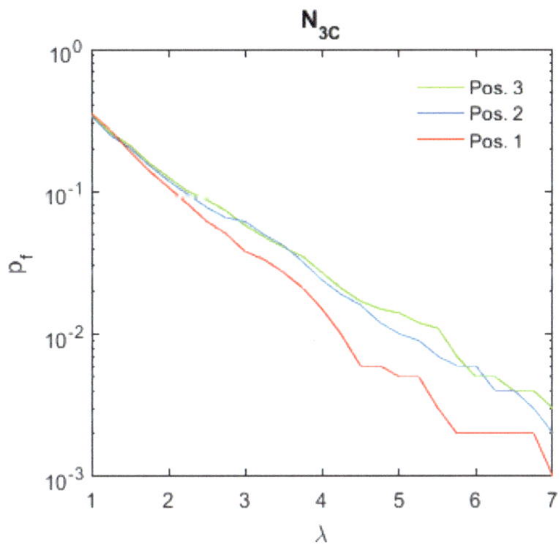

Figure 21.
Failure probability for N_{3C} for different λ values.

Regarding p_f as the failure probability of the structural element, $N_{G<0}$ is the number of simulations in which a performance function for the number of cycles to failure for the #C compliance $G_{N_{3C}} < 0$:

$$p_f = \frac{N_{G<0}}{N_{total}} \tag{4}$$

$$G_{N_{3C}} = \underbrace{\lambda \mu_{N_{3C}}}_{\substack{resistant \\ fatigue \\ life}} - \underbrace{N_{3C}}_{\substack{loading \\ fatigue \\ life}} \tag{5}$$

In which N_{total} the total number of simulations, λ is like an inverse safety factor that scales an allowable (resistant) fatigue life, and $\mu_{N_{3C}}$ the loading average life. **Figure 21** shows the probability of failure of the three positions and different values of λ. It is noticed that the failure probability is greater for Position 3, followed by Position 2 and later from Position 1.

5. Conclusions

Due to the flexibility of the Boundary Elements Method, it was possible to develop the damage tolerance technique by assessing the compliance of the micro-element edges on an aircraft fuselage metal plate. The automation of the thousands of Monte Carlo simulations using an in-house BEM computer program enabled the probabilistic analysis of the fatigue life of the aircraft fuselage in which the plate may have initial defects. The analysis carried out the influence of random input variables of material properties and loadings on the fatigue life predicted values. Based on this technique, the study showed that when the initial defects occur at the edge of the external loading application position for low numbers of cycles, the microelement has high compliance tending to sudden instability, being, therefore, the most unfavorable case. It was possible to observe that when compliance reaches the value of 3 times the initial compliance, it leads to local instability of the microelement. With the data series, it was possible to perform a statistical treatment to define the damage tolerance to avoid the occurrence of a Limit State. In general, the use of this methodology is shown as an alternative to the analysis of damage tolerance regarding the process adopted by the literature in which the damage is considered from a critical crack size. In the method employed, critical size is disregarded and compliance is assessed as a variable that defines instability.

Acknowledgements

The authors thank the Graduate Programme in Structural Engineering and Civil Construction at the University of Brasilia, Coordination for the Improvement of Higher Education Personnel (CAPES) and National Council for Scientific and Technological Development (CNPq).

Author details

Gilberto Gomes*, Thiago Oliveira and Francisco Evangelista Jr
University of Brasilia, Brasilia, Brazil

*Address all correspondence to: ggomes@unb.br

References

[1] National Transportation Safety Board - NTSB, "Aircraft Accident Report, Aloha Airlines, Flight 243, Boeing 737-200, N73711, Near Maui, Hawaii, April 28, 1988," Washington, DC, 1989.

[2] R. J. H. Wanhill, "Milestone case histories in aricraft structural integrity," *Comprehensive structural integrity,* vol. 1, no. I. Milne, R. O. Ritchie, B. Karihaloo, pp. 61-72, 2003.

[3] J. Jisan and Y. Xiaochuan, "Dynamic fracture analysis technique of aircraft fuselage containing damage subjected to blast," *Mathematical and Computer Modelling. Elsevier,* 2011.

[4] A. Portela, M. H. Aliabadi and D. P. Rooke, "Dual boundary element analysis of cracked plates: singularity subtraction technique," *International Journal of Fracture,* vol. 55, pp. 17-28, 1993.

[5] G. E. Blandford, A. R. Ingraffea and J. A. Liggett, "Two-Dimensional Stress Intensity Factor Computations Using the Boundary Element Method," *International Journal Numerical Methods in Engineering,* no. 17, pp. 387-404, 1981.

[6] A. Portela, M. H. Aliabadi and D. P. Rooke, "The Dual Boundary Element Method: Efective implementation for crack problems," *International Journal for Numerical Methods in Engineering,* vol. 33, pp. 1269-1287, 1992.

[7] R. Citarella et al., "DBEM crack propagation in friction stir welded aluminum joints," *Advances in Engineering Software,* vol. 101, pp. 67-75, 2016.

[8] R. Citarella, "MSD crack propagation by DBEM on a repaired aeronautic panel," *Advances in Engineering Software,* vol. 42, no. 10, pp. 887-901, 2011.

[9] R. J. Price and J. Trevelyan, "Boundary element simulation of fatigue crack growth in multi-site damage," *Engineering Analysis with Boundary Element,* vol. 43, pp. 67-75, 2014.

[10] L. Morse , S. Khodaei and M. H. Aliabadi, "Multi-Fidelity Modeling-Based Structural Reliability Analysis with the Boundary Element Method," *Journal of Mulstiscale Modelling,* vol. 08, 2017.

[11] X. Huang, M. H. Aliabadi and Z. S. Khodaei, "Fatigue Crack Growth Reliability Analysis by Stochastic Boundary Element Method," *CMES,* vol. 102, pp. 291-330, 2014.

[12] B. Palmberg, A. F. Blom and S. Eggwertz, "Probabilistic damage tolerance analysis of aircraft structures," *Engineering Application of Fracture Mechanics,* vol. 6, no. Probabilistic fracture mechanics and reliability, pp. 47-130, 1987.

[13] R. J. H. Wanhill, "Flight simulation fatigue crack growth testing of aluminium alloys: Specific issues and guidelines," *International Journal of Fatigue,* vol. 16, no. 2, pp. 99-110, 1994.

[14] R. J. H. Wanhill, "Status and prospects for aluminium-lithium alloys in aircraft structures," *International Journal of Fatigue,* vol. 16, no. 1, pp. 3-20, 1994.

[15] J. C. Newman Jr., "The Merging of Fatigue and Fracture Mechanics Concepts: A Historical Perspective," *Fatigue and Fracture Mechanics,* vol. 28, no. J. H. Underwood, B. D. MacDonald, and M. R. Mitchell, 1997.

[16] J. C. Newman Jr., E. P. Phillips and M. H. Swain, "Fatigue-life prediction methodology using small-crack theory," *International Journal of Fatigue,* no. 21, 1999.

[17] J. Schijve, "Fatigue damage in aircraft structures, not wanted, but to be tolerated?," in *International Conference on Damage Tolerance of Aircraft Structure*, Delft, 2007.

[18] S. Barter, L. Molent, N. Goldsmith and R. Jones, "An experimental evaluation of fatigue crack growth," *Engineering failure analysis,* no. 12, pp. 99-128, 2005.

[19] S. A. Barter, L. Molent and R. J. H. Wanhil, "Typical fatigue-initiating discontinuities in metallic aircraft structures," *International Journal of Fatigue,* vol. 41, pp. 11-22, 2012.

[20] L. Molent and C. Forrester, "The lead crack concept applied to defect growth in aircraft composite structures," *Composite Structures,* vol. 166, pp. 22-26, 2017.

[21] P. Chowdhury, H. Sehitoglu and R. Rateick, "Damage tolerance of carbon-carbon composites in aerospace application," *Carbon,* 2017.

[22] N. M. Chowdhurry, W. K. Chiu, J. Wang and P. Channg, "Experimental and finite element studies of bolted, bonded and hybrid step lap joints of thick carbon fibre/epoxy panels used in aircraft structures," *Composites,* vol. 100, pp. 68-77, 2016.

[23] F. Pegorin, K. Pingkarawat and A. P. Mouritz, "Comparative study of the mode I and mode II delamination fatigue properties of z-pinned aircraft composites," *Materials & Design,* vol. 65, pp. 139-146, 2015.

[24] L. Smith, R. Pilarczyk and J. Feiger, "Validation Testing and Analysis of Cracked-Hole Continuing Damage Solutions," *Materials Performance and Characterization,* vol. 5, no. 3, 2016.

[25] F. O. Sonmez, "Shape optimization of 2D structures using simulated annealing," *Computer Methods in Applied Mechanics and Engineering,* vol. 196, no. 35-36, pp. 3279-3299, 2007.

[26] S. Chintapalli, M. S. A. Elsayed, R. Sedaghati and M. Abdo , "The development of a preliminary structural design optimization method of an aircraft wing-box skin-stringer panels," *Aerospace Science and Technology,* vol. 14, no. 3, pp. 188-198, 2010.

[27] M. Gorelik, "Additive manufacturing in the context of structural integrity," *International Journal of Fatigue,* vol. 94, no. 2, pp. 168-177, 2017.

[28] M. Ciavarella and A. Papangelo, "On the distribution and scatter of fatigue lives obtained by integration of crack growth curves: Does initial crack size distribution matter?," *Engineering Fracture Mechanics,* vol. 191, pp. 111-124, 2018.

[29] A. K. Srivastava, P. K. Arora and H. Kumar, "Numerical and experiment fracture modeling for multiple cracks of a finite aluminum plate," *International Journal of Mechanical Sciences,* vol. 110, pp. 1-13, 2016.

[30] F. Evangelista Jr., J. Roesler and C. Duarte, "Two-Scale Approach to Predict Multi-Site Cracking Potential in 3-D Structures Using the Generalized Finite Element Method.," *International Journal of Solids and Structures,* vol. 50, no. 1, pp. 1991 - 2002, 2013.

[31] P. H. Wen, M. H. Aliabadi and A. Young, "Crack growth analysis for multi-layered airframe structures by boundary element method," *Engineering Fracture Mechanics,* vol. 71, no. 4-6, pp. 619-631, 2004.

[32] N. K. Salgado and M. H. Aliabadi, "The analysis of mechanically fastened repairs and lap joints," *Fatigue & Fracture of Engineering Materials & Structures,* vol. 20, no. 4, pp. 583-593, 1997.

[33] P. C. Paris and F. Erdogan, "A critical analysis of crack propagation laws," *Journal of basic engineering,* no. 85, pp. 528-534, 1960.

[34] K. J. Bathe and E. L. Wilson, Numerical methods in finite element analysis, New Jersey: Prentice Hall, Inc., 1976.

[35] C. A. Brebbia, The boundary element methods for engineers, New York: Halstead Press, 1978.

[36] A. R. Ingraffea, G. E. Blandford and J. A. Ligget, "Automatic Modelling of Mixed-Mode Fatigue and Quasi-Static Crack Propagation Using the Boundary Element Method,," in *Proc. of Fracture Mechanics: Fourteenth Symposium, ASTM STP 791*, 1983.

[37] G. Gomes and A. C. O. Miranda, "Analysis of crack growth problems using the object-oriented program BemCracker2D," *Frattura ed Integrità Strutturale,* vol. 45, pp. 67-85, 2018.

[38] G. Gomes, A. M. Delgado Neto and L. C. Wrobel, "Modelling and 2D cracks view using dual boundary integral equation.," in *XXXCII Iberian Latin American Congress on Computational Methods in Engineering - CILAMCE*, Brasília, 2016.

[39] T. A. A. Oliveira , G. Gomes and F. Evangelista Jr., "Multiscale aircraft fuselage fatigue analysis by the dual boundary element method," *Engineering Analysis with Boundary Elements,* no. 104, pp. 107-119, 2019.

[40] A. M. Delgado Neto, G. Gomes and T. A. A. Oliveira, "An efficient GUI update for BEM-FEM mixed mesh generation," *International Journal for Computational Methods in Engineering Science and Mechanics,* pp. 256-267, 2019.

[41] P. G. P. Leite and G. Gomes, "Numerical simulation of fatigue crack propagation in mixed-mode (I+II) using the program BemCracker2D," *International Journal of Structural Integrity,* vol. 10, no. 4, pp. 497-514, 2019.

[42] M. H. Aliabadi, The Boundary Element Method - Application in Solids and Structures, vol. 2, Wiley, 2002.

[43] N. D. e. a. Alexopoulos, "The effect of artiĄcial ageing heat treatments on the corrosion-induced hydrogen embrittlement of 2024 (alŰcu) aluminium alloy.," *Corrosion Science,* vol. 102, pp. 413-424, 2016.

[44] Y. e. a. Lin, "Precipitation hardening of 2024-t3 aluminum alloy during creep aging.," *Materials Science and Engineering: A, Elsevier,* vol. 565, pp. 420-429, 2013.

[45] N. E. Dowling, Mechanical Behavior of Materials, 4 ed., S. C.: Prentice Hall, 2012.

[46] D. Broek and J. Schijve, "The influence of the mean stress on the propagation of fatigue cracks in aluminium alloy sheet," in *NLR-TR M2111*, 1963.

Chapter 4

Combination of Numerical, Experimental and Digital Image Correlation for Mechanical Characterization of Al_2O_3/β-TCP Based on CDM Criterion

Barkallah Rachida, Rym Taktak, Noamen Guermazi, Fahmi Zaïri and Jamel Bouaziz

Abstract

Cracks in engineering materials and structures can undergo different modes of deformation. This chapter presents a numerical and experimental approaches aimed to assess the fracture toughness and the Fracture behavior under tensile and shear loading of bioceramics based on commercial Alumina (Al_2O_3), synthesized Tricalcium phosphate (β-TCP). Conditioning was conducted at different percentages of TCP. After a sintering process at 1600°C for 1 hour, The Crack Straight Through Brazilian Disc were performed by image correlation during a mechanical test and numerical tests were carried out in order to find the angle where the pure mode II. A CDM based constitutive model was selected and implemented into a finite element code to study the damage of our bioceramics. The result of this combination was compared with the direction of crack propagation obtained experimentally. The directions of crack propagation found numerically were found in good agreement with those experimentally obtained by a mechanical test. Alumina-10 wt.% Tricalcium phosphate composites displayed the highest values of the fracture toughness. This value reached 8.76 MPa $m^{1/2}$ MPa. The same optimal composition for the mode I and mode II stress intensity factor with maximum values of 7.6 MPa $m^{1/2}$ and 8.45 MPa $m^{1/2}$ respectively.

Keywords: Fracture, Modeling, Tensile loading, Shear loading

1. Introduction

Recently, the technology of tissue engineering has widely known in substantial advancements and innovations. This technology is a discipline to restoring the task of various organs through the regeneration and also develop novel synthetic biomaterials. It is being investigated and applied in most organ systems, restoring the function of various tissues and organs, such as heart valves, blood vessels and orthopedic implants, among many others [1–3]. Cracks and flaws which certainly exist in the sample reduce in a significant way the load-bearing capacity and then cause the substitute to break [4, 5]. The fracture toughness and stress intensity factor have been proposed to express the critical stress states in the vicinity of the crack tip, in the aim to analyze crack initiation and propagation [5].

Calcium phosphate bioceramics, with its excellent biological properties, such as biocompatibility and osteconductivity and its outstanding mechanical properties, including hardness, low density and its inertness at high temperature, is widely known as a suitable candidate for biomaterials. Despite their advantages, Calcium phosphates bioceramics exhibit very low toughness which limits their overall applications [6]. The challenge of increasing the toughness of bioceramic has been a key motivation in the field of biomaterials research. In this pursuit of improving toughness, β-tricalcium phosphate (β-Ca3(PO4)2) (β-TCP) are often used due to its outstanding biological responses to physiological environments [7]. The introduction of Alumina (Al2O3) toughening agent increased the toughness of the tricalcium phosphate composite.

Alumina has been widely studied due to its high wear resistance, fracture toughness and strength as well as relatively low friction without forgetting its bioinertness [8].

In recent investigation, Barkallah et al. [9] have been concerned with the Alumina - Tricalcium phosphate composites with different percentages. These Al_2O_3/β-TCP composites have shown a good combination of elastic modulus (76 GPa), tensile strength (27 MPa), compressive (173 MPa) and flexural strength (66 MPa) but this biomaterial has never been investigated the stress intensity factors in Crack Straight Through Brazilian Disc specimen, under tensile and shear loading and their crack's initiation and its propagation. Those parameters of the developed composites should be evaluated.

In fact, there are three basic fracture propagation modes (**Figure 1**): Mode I (opening mode), Mode II (in-plane shear mode), and mixed mode [10]. In pure mode I loading, any two respective points along the notch faces open relative to the notch bi-sector line without any sliding. Under pure mode II, the two respective points along the notch faces slide relative to the notch bi-sector line without any opening and the tangential stress along the bi-sector line is zero. Any combination of mode I and mode II deformation is called mixed mode loading. The shear stress along the bi-sector line is zero for only the loading is pure mode I [11].

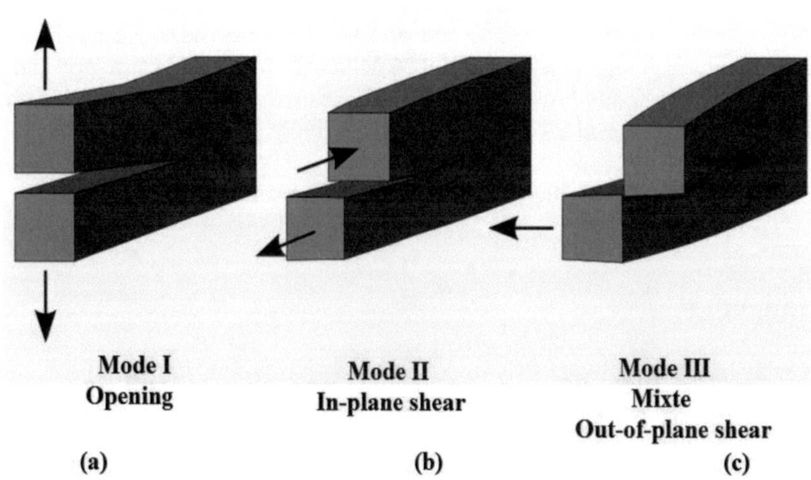

Figure 1.
Basic modes known in fracture mechanics: (a) tensile opening (mode I), (b) In-plane shear (mode II) and (c) out-of-plane shear (mode III).

Many different test specimens have been proposed in the past for brittle or quasi-brittle materials for determining the mode I, II fracture toughness for various engineering materials [12–14]. The centrally cracked Brazilian disc specimen has been used by many researchers to study mode I and mode II fracture mechanics in different brittle materials [11, 14].

Because of the brittleness of Biomaterials based on ceramics, the study of the contact problem with external objects is important. However, ceramics and bioceramics are inherently brittle. This characteristic leads, in particular, to a wide variation in the material strength. A CDM based constitutive model have been developed to study the damage of our bioceramics and thanks to this model, the numerical modeling of the damage behavior of bioceramics during a mechanical test is reported. This modeling is essential for a better understanding of fracture mechanisms of bioceramics [15].

On the other hand, in the last 20 years, digital image correlation (DIC) has shown that it is a valuable non-contact technique for measuring kinematic fields during a mechanical test [16, 17]. In order to account for the maximum load, it is crucial to work with local displacements at the damage progress zone.

In this chapter, we present a damage model in combination with finite element technique that can help automatize the damage progress fracture in an efficient manner. Our work was undertaken to evaluate the mechanical behavior of the combination of commercial alumina with synthetic Tricalcium phosphate as bone substitute material. To achieve this purpose, we study the stress intensity factor KI under tensile stress (mode I rupture) and stress intensity factor KII under shear stress (mode II rupture experimentally and theoretically using modified Brazilian test. The samples were also characterized by scanning electron microscopy (SEM).

2. Materials and methods

2.1 Materials

In order to elaborate Al_2O_3-TCP, the materials used were commercial Al_2O_3 (Riedel-de Haën, purity >98%) and synthesized tricalcium phosphate powders.

The β-TCP powder was synthesized by solid-state reaction from calcium phosphate dibasic anhydrous ($CaHPO_4$) and calcium carbonate ($CaCO_3$). Stoichiometric amounts of high purity powders, $CaCO_3$ (Fluka, purity ≥98.5%) and $CaHPO_4$ (Fluka, purity ≥99%) were sintered at 1000°C for 20 hours to obtain the β-TCP according to the following reaction [18]:

Synthesized Tricalcium phosphate was introduced with α-Alumina powders. The approximate representatives Al_2O_3-TCP were {90 wt.%, 10 wt.%}, {80 wt.%, 20 wt.%}, {60 wt.%, 40 wt.%} and {50 wt.%, 50 wt.%}.

It is worth mentioning that the size of particles of each powder was measured (2.53 μm for Al_2O_3 and 2.79 μm for TCP).

As starting materials, calculated quantities of The β-TCP and Al_2O_3 powder were mixed by homogeneous mixing in a mortar and milled in absolute ethanol and treated with an ultrasound machine for 20 min. After milling these powders, the mixture was dried at 80°C for 24 hours to eliminate the ethanol. After drying, mixtures powders thus prepared were molded in a metallic cylinder mold having a diameter of 30 mm and a thickness of 5 mm and uniaxially pressed at 67 MPa. A crack "2a" of 12 mm was added to the CSTBD specimens were considered for each compacted specimens. The crack of the CSTBD specimens was added by a specific metal mold (**Figure 1**). At least six specimens were tested under each test condition.

Finally, the specimens were sintered at 1600°C for 1 hour in a vertical programmable muffle furnace (Pyrox 2408) and were heated and cooled at a rate of 10° C min^{-1} and 20°C min^{-1}, respectively.

On the other hand, one side of each fracture sample has been sprayed with black then white paint in order to form the speckle pattern that will be used by the Digital Image Correlation (DIC) technique.

2.2 Experimental part

At least six specimens were tested under each condition and then average values (K_{IC}, K_I and K_{II}) were considered.

2.2.1 Bending test.

The fracture toughness K_{IC} of the samples were assessed using Semi Circular Bending tests. The samples were positioned on the loading platform by 3-point compressive loading, at a uniform loading speed of 0.075 mm/min (**Figure 2a**). The SCB specimen diameter is equal to 30 mm and 5 mm for thickness. The specimen contains a crack of 4 mm in the semi disc, as shown in **Figure 2b**. The crack-length-to diameter ratio S/D was 0.13.

Using the SCB specimen with straight crack, the fracture toughness K_{IC} was calculated with the following formula [19]:

$$K_{IC} = \frac{P_{\max}\sqrt{\pi a}}{2Rt} Y_I(a/R, S/R) \quad (1)$$

Where a is the crack length, P_{max} is the maximum load, D is the cylindrical block diameter and YI is the geometry factor. The latter is a function of the ratio of the crack length (a) over the semi-disc radius (R) and the ratio of the half-distance between the two bottom supports (S) over the semi-disc radius (R) (**Figure 2b**). The geometry factor Y_I is expressed as follows [19]:

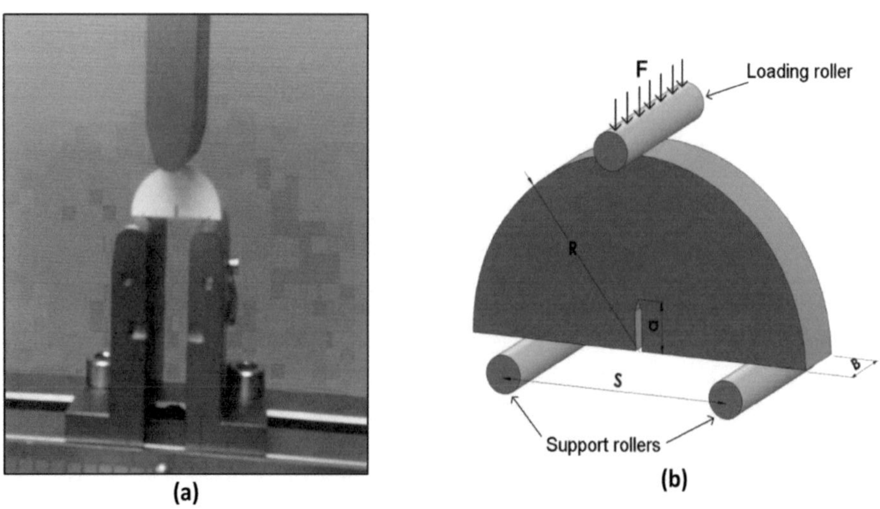

Figure 2.
Semi-circular bending (SCB): (a) real photo of SCB test (b) illustration of cracked SCB specimen.

$$Y_I(a/R, S/R) = \frac{S}{R}\left(2.91 + 54.39\frac{a}{R} + 391.4\left(\frac{a}{R}\right)^2 + 1210.6\left(\frac{a}{R}\right)^3 - 1650\left(\frac{a}{R}\right)^4 + 875\left(\frac{a}{R}\right)^5\right) \tag{2}$$

2.2.2 Cracked straight-through Brazilian disc

The introduction of the fracture mechanics approach to brittle materials has led to the development of materials fracture mechanics, which refers to the initiation and propagation of a crack or many cracks in materials.

According to the applied stress condition, a crack propagates depending on the three basic failure modes [20]: Mode I loading state is defined as opening mode, the mode II is defined as sliding mode (shear mode) and mode III is defined as tearing.

In this chapter, only mode I and II will be studied and detailed. Bioceramic stress intensity factor under modes I and II was measured using CSTBD specimens for an experimental and analytical investigation [21, 22]. Disc-type specimens are simple in geometry and have many advantages in terms of sample preparation, testing and analysis.

Different combinations of mode I and mode II can be shown by changing the crack angle β: if the direction of compressive applied load is along the crack bi-sector line β = 0, the samples is subjected to pure mode I loading. If β ≠ 0, the samples are subjected to mixed mode I/II loading. A gradual increase of the loading angle results in an elevation in mode II effects and reduction in mode I effects. Finally, there are a specific loading angle β_II for which the sample undergoes pure mode II deformation. This angle was found in this research by a series of finite element analyses [11].

The UMTS criterion is a criterion for brittle fracture is proposed by Ayatollahi [11] for prediction the mode II fracture toughness of U notched components and the fracture initiation angle in CSTBD under pure mode II loading.

The International Society for Rock Mechanics (ISRM) proposed many analytical formulas for measuring fracture toughness mode I of brittle materials: the cracked chevron notched Brazilian disc (CCNBD) specimens and the Cracked straight through Brazilian disc (CSTBD) [12, 23] (see **Figure 3a** and **b**). The CSTBD and CCNBD specimens has the same geometry and shape as the conventional Brazilian disc used for measuring the indirect tensile strength, except that the CSTBD specimen has a through notch length of 2a, by means of the straight-through crack assumption (STCA) method.

By comparing these two methods, CSTBD has superiority over CCNBD considering that producing a stream crack is easier than a V-shape crack. [14, 24]

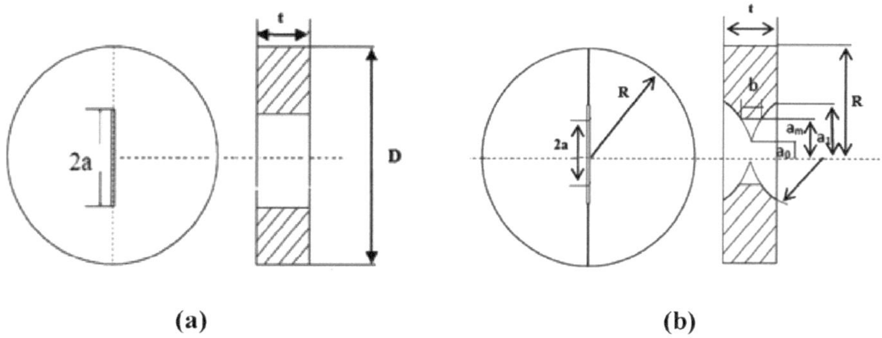

Figure 3.
Disc-type specimens: (a) CSTBD and (b) CCNBD.

The stress intensity factor (SIF) solutions for CSTBD specimens can be met in Cherepanov's book [25] and the handbook [26]. the main formulas to remember are collected as follows:

Cherepanov's book:

$$K_I = \frac{P}{\sqrt{D}\,t} Y \tag{3}$$

$$Y = \sqrt{\frac{2}{\pi}}\sqrt{\alpha}\left[1 + \frac{3}{2}\alpha^2 + \frac{3}{4}\alpha^6 + \frac{3}{64}\alpha^8\right] \tag{4}$$

The handbook:

$$K_I = \frac{P}{\sqrt{D}\,t} Y \tag{5}$$

$$Y = \sqrt{\frac{2}{\pi}}\sqrt{\frac{\alpha}{1-\alpha}}\left[1 - 0.4964\,\alpha + 1.5582\,\alpha^2 - 3.1818\,\alpha^3 + 10.096\,\alpha^4 - 20.7782\,\alpha^5 + 20.1342\,\alpha^6 - 7.5067\,\alpha^7\right] \tag{6}$$

where $\alpha = \frac{a}{R}$ knowing that K_I is the mode-I stress intensity factor, Y is the dimensionless stress intensity factor, P is the concentrated diametral compressive load, D is the diameter, t is the thickness and a is the crack length.

According to [27], the analytical solution of the stress intensity factor SIF for the CSTBD specimen for measuring the fracture toughness of ceramics can be expressed in the following form Shetty et al. [28]:

$$K_I = \frac{P}{\pi R t}\sqrt{\pi a}N_I = \frac{P}{\sqrt{\pi}\,R t}\sqrt{a}N_I \tag{7}$$

Where P is the load applied in compression, a is half the notch length and N_I is the dimensionless stress intensity factor depending on the dimension less crack length α (a/R) and the notch inclination β.

N_I solutions for the CSTBD sample can be determined by several methods:

Starting with Atkinson et al. [29] who has developed N_I solutions for determining the fracture toughness and applied the stress intensity factor solutions of the CSTBD.

By small crack approximation ($\alpha \le 0.3$) and five-term approximation, N_I was developed as the following formula:

$$\boldsymbol{N_I} = 1 - 4\,sin^2\beta + 4\,sin^2\beta(1 - 4\,cos^2\theta)\alpha^2 \tag{8}$$

With

$$\alpha = \frac{a}{R}$$

when β = 0, the problem is reduced to theMode I fracture situation, then according to, the previously equation $\boldsymbol{N_I}$ becomes [28]:

$$\boldsymbol{N_I} = 0.991 + 0.141\,\alpha + 0.863\,\alpha^2 + 0.886\,\alpha^3 \tag{9}$$

Wherever, Fowell et al. [22] developed the formula on other form $0.05 \leq \alpha \leq 0.95$:

$$\begin{aligned} N_I = \sqrt{\frac{\pi}{\alpha}}(&0.0354 + 2.0394\,\alpha - 7.0356\,\alpha^2 + 12.1854\,\alpha^3 + 8.4111\,\alpha^4 - 30.7418\alpha^5 \\ &- 29.4959\,\alpha^6 + 62.9739\,\alpha^7 + 66.5439\,\alpha^8 - 82.1339\,\alpha^9 - 73.6742\,\alpha^{10} \\ &+ 73.8466\,\alpha^{11}) \end{aligned} \tag{10}$$

As mentioned, different combinations of mode fracture can be obtained by changing the angle β. For while mode II, we can find the specific loading angle β, for which the specimen undergoes pure mode II deformation, by a series of finite element analyses. The mode II loading angle β was then determined from finite element results for the notch length that is already selected for mode I.

The stress intensity factor, for the CSTBD specimen with a through notch length of 2a, under mode II can be calculated with the following formula [28]:

$$K_{II} = \frac{P}{\pi R t}\sqrt{\pi a}N_{II} = \frac{P}{\sqrt{\pi} R t}\sqrt{\alpha}N_{II} \tag{11}$$

$$N_{II} = \left[2 + \left(8\cos^2\theta - 5\right)\alpha^2\right]\sin 2\beta \tag{12}$$

Where P is the load applied in compression, a is half the crack length and N_{II} is the dimensionless stress intensity factor under mode II, depending on the dimensionless notch length α=a/R and the crack inclination angle with respect to loading direction, β (**Figure 4**).

In this case, N_{II} was developed by [29].

2.2.3 Digital correlation

Nowadays, various full-field non-contact optical methods have been reported in literature and succeeded in replacing those classical techniques by Digital image correlation for strain and displacement measurements [16, 30]. The principle of DIC analysis is based on the comparison of the different successive digital images acquired during the test.

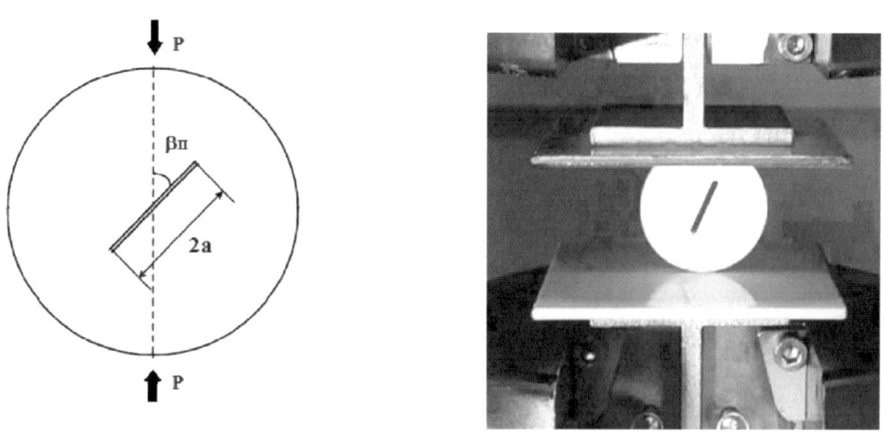

Figure 4.
CSTBD under pure mode II fracture.

As mentioned in introduction, the experimental displacement was here computed by using Digital Image Correlation (DIC) in order to determinate the crack propagation at different states of loading and different composition. In the present work, DIC calculations have been managed with Correla software.

At each load step and at each composition, a series of images is taken with a CCD camera and digitalized and then compared to the reference image. For this technique, the displacement field analysis was performed inside of a Region of Interest (ROI) divided into discrete subsets. The shape (square or rectangular), the size (number of pixels) and the distribution (vertical and horizontal distances between centres, (Lx,Ly)) of these subsets should be carefully chosen. Those parameters depending to the desired accuracy of measurements (displacement and strain) and to the spatial resolution for map fields [16].

For each subset, a correlation function is used to estimate the degree of similarity between the reference image state and the current one (for each given load) [31].

Increasing the sub-set size allows decreasing the uncertainty because DIC error mainly depends on the number of pixels in thesubset [17]. The dedicated subsets were voluntary chosen with different scale factor (as shown on **Table 1**).

2.3 CDM model

In this study, the FE simulations were performed by a constitutive model describing the mechanical behavior of brittle material is based on the CDM approach.

At first, The CDM approach was introduced by Kachanov [32] and generalized later by Le maitre and Chaboche [33]. It is a concept which provides a mathematical description of the effect of micro-defects and micro-cracks, at a macro-scale, on the macroscopic properties of the material. After that, the works of J. Ismail [15] show that even the mathematical formulation of this damage mechanics model that allows it to predict the cracking damage patterns in brittle materials.

Composition	Scalor factor (mm/pixel)
Alumine	0,047138
Al_2O_3–10 wt.% TCP	0,047138
Al_2O_3–20 wt.% TCP	0,0474517
Al_2O_3–40 wt.% TCP	0,0470807
Al_2O_3–50 wt.% TCP	0,0480663
Al_2O_3–10 wt.% TCP - 1 wt.% TiO_2	0,04803
Al_2O_3–10 wt.% TCP - 2,5 wt.% TiO_2	0,04803
Al_2O_3–10 wt.% TCP - 3 wt.% TiO_2	0,0481132
Al_2O_3–10 wt.% TCP - 4 wt.% TiO_2	0,048955
Al_2O_3–10 wt.% TCP - 5 wt.% TiO_2	0,0480843
Al_2O_3–10 wt.% TCP - 7,5 wt.% TiO_2	0,0481766
Al_2O_3–10 wt.% TCP - 10 wt.% TiO_2	0,0481766

Table 1.
Calibration scalor factor.

For an exact estimation of damage patterns, there is a clear need of constitutive equation of brittle material that is defined by:

$$\sigma = k * \varepsilon \tag{13}$$

where K is the fourth-order stiffness tensor which is written as

$$k = k^e + k^d \tag{14}$$

in which k^e denotes the fourth-order stiffness tensor for the isotropic virgin material. k^d is a fourth-order tensor which represents the added damage influence and the final expression is given by

$$k^d_{ijkl} = C1\left(\delta_{ij}D_{kl} + \delta_{kl}D_{ij}\right) + C2\left(\delta_{jk}D_{il} + \delta_{il}D_{jk}\right) \tag{15}$$

where δ is the Kronecker-delta symbol and C1 and C2 are the damage parameters.

The damage variables that expressed as functions of stress state is introduced into an anisotropic damage tensor D_{ij}. Their values vary between 0 for virgin state and 1 for fully damaged (cracking) state.

Both damage patterns (mode I/mode II) are modeled by taking into consideration the effects of tensile principal stresses as well as compressive and shear stresses.

For the functioning of damage tensor, account should be taken of the effect of normal principal stress and those shear stress components in the damage mechanisms in brittle materials and the damage for both modes I and II is modeled.

- The first types of components (mode I)

This mode is involved by normal tensile principal stresses. The damage compo nents representing are the diagonal terms of tensor K and their values are expressed in term of critical and threshold stress limits by:

.1

$$D_{ii} = \begin{cases} 0 \; if \; \sigma_i \leq \sigma_t \\ \dfrac{\sigma_i - \sigma_t}{\sigma_c - \sigma_t} \; if \; \sigma_t < \sigma_i < \sigma_c \quad i = 1, 2, 3 \\ 1 \; if \; \sigma_i \geq \sigma_c \end{cases} \tag{16}$$

Where σ_c and σ_t are the critical and threshold stresses that corresponds to the stress below which no damage occurs.

- The second types of components (mode II)

In some cases, the shear mode (mode II) could be activated. Those components are formulated as a function of shear stress in the symmetry plane. The general form is:

$$D_{ij} = \begin{cases} 0 \quad \textit{if } \sigma_{ij} \le \tau_t \textit{ and } \max(\sigma_i) > 0 \\ \dfrac{\sigma_{ij} - \tau_t}{\tau_c - \tau_t} \quad \textit{if } \tau_t < \sigma_{ij} < \tau_c \textit{ and } \max(\sigma_i) < 0 \\ 1 \quad \textit{if } \sigma_{ij} \ge \tau_c \textit{ and } \max(\sigma_i) > 0 \end{cases} \tag{17}$$

$$i = 1, 2, 3 \textit{ and } i \ne j$$

Where τ_c and τ_t are the critical and threshold shear parameters, respectively. The damaged constitutive equations were coded in the Fortran programming language and implemented in the commercial FE code MSC. Marc to simulate the behavior and damage evolution in our materials.

The micro-cracks and micro-defects are an irreversible phenomenon. For this reason, the damage does not decrease during the loading and the D_{ij} is taken as a monotonic increasing function of time increment:

$$D_{ij} = \max\left(D_{ij}^{n}, D_{ij}^{n-1}\right) \tag{18}$$

Where D_{ij}^{n} is the damage value at the current time step n and D_{ij}^{n-1} is the damage value at the previous time increment n-1.

A procedure to identify the model parameters must now be defined. The input parameters required are:

- The Poisson's ratio
- The Young's modulus
- The critical and threshold stresses: c and t in mode I.
- The breaking load

The material properties such as the ultimate tensile strength (σ_t), the elastic modulus (E) and the Poisson's ratio (μ) for Al_2O_3-TCP were determined experimentally using the standard test techniques at room temperature.

2.4 Finite element analysis

In this section, the commercial FE code MSC. Marc was used to perform the simulations. A two-dimensional calculation has been performed using the finite element program MARC. A plane stress FE model with a total number of 8000 Quad 4 elements was created for simulating the specimen by moving two plates to effect compression on the disk. **Figure 5a** shows a sample FE grid pattern used for simulating a CSTBD specimen. The finest elements were located near the notch tip due to its high stress gradient (**Figure 5b**).

In order to determine the angle β_{II} for which the sample undergoes pure mode II deformation, this angle was found in this research by a series of finite element analyses. Then, the values of the tangential and the shear stresses $(\sigma_{12}, \sigma_{22})$ along the notch bi-sector line could be obtained from the FE results In a Cartesian coordinate. In an auxiliary system of curvilinear coordinates, when $\sigma_{\theta\theta}(r_0, 0) = 0$, the mode I is zero, and hence the specimen is subjected to pure mode II deformation. Therefore, the mode II loading angle β_{II} is the angle for which $\sigma_{\theta\theta}(r_0, 0) = 0$ [11].

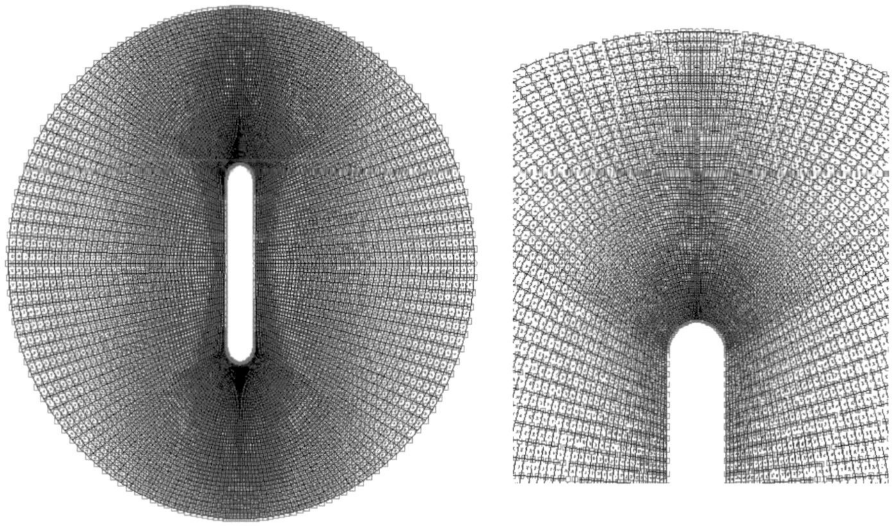

Figure 5.
A FE grid used for the simulations: (a) FE-mesh for the whole sample, (b) FE-mesh near the notch tip.

After having used the matrix for passing from a Cartesian coordinate system to a cylindrical coordinate system, $\sigma_{\theta\theta}$ can write Eq. (19),

$$\sigma_{\theta\theta} = \sigma_{22} * \cos(\theta) - \sin(\theta) * \sigma_{12} \tag{19}$$

In order to obtain the pure mode II loading angle β_{II}, the angle b was gradually increased from zero and the value of tangential stress $\sigma_{\theta\theta}(r_0, 0)$ at the notch tip was calculated for each loading angle, under a compressive load already found by the mechanical tests. As the loading angle increased, the value of $\sigma_{\theta\theta}(r_0, 0)$ decreased until it was equal to zero.

In a second step, a CDM criterion for brittle materials has been introduced in the MARC-2005 to predict the mechanical behavior of our biomaterials subjected to a mechanicals test and this modeling was used to simulate the damage process. This combination has allowed to detect crack initiation and to analyze fracture process.

The mechanical properties were chosen to represent the composite specimens, for which elastic modulus and Poisson's ratio are (47.03; 75.96; 55.75; 46.86 and 33.51 GPa) and (0.283; 0.318; 0.361; 0.363 and 0.28), respectively for the variation of TCP.

3. Results and discussion

3.1 Determination of β under mode II

The CSTBD specimen has been used by many researchers to study mode I and mode II brittle fracture in different materials. However, the experimental results obtained in the past part from this specimen indicate that the mode II is β = 22° [34] and β = 25° [11].

Depending on the crack length and the disc radius already chosen, the numerical calculation reported in this figure (**Figure 6**) shows that the stress (σ) is equal to zero near to angle 22, this result is similar for all compositions, so this angle verifies the pure mode II. Therefore, the results of the FE analysis have been obtained shows

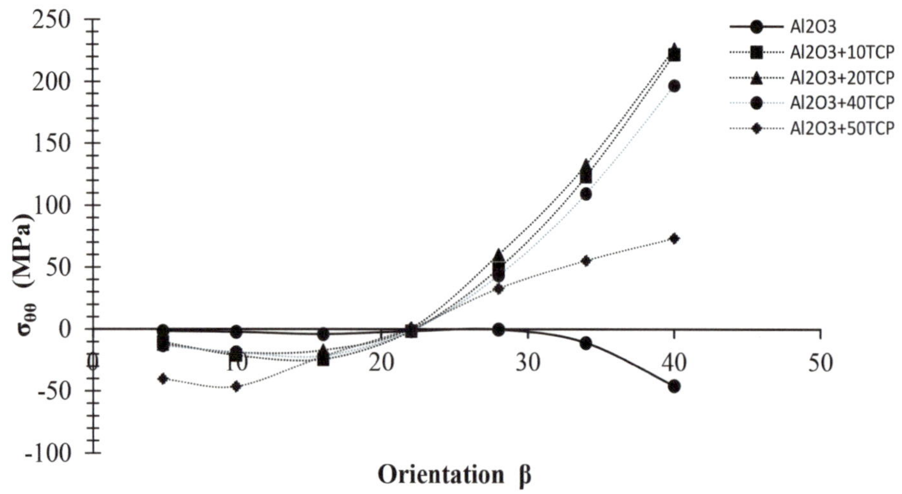

Figure 6.
Variation of $\sigma_{\theta\theta}$ for each compositions.

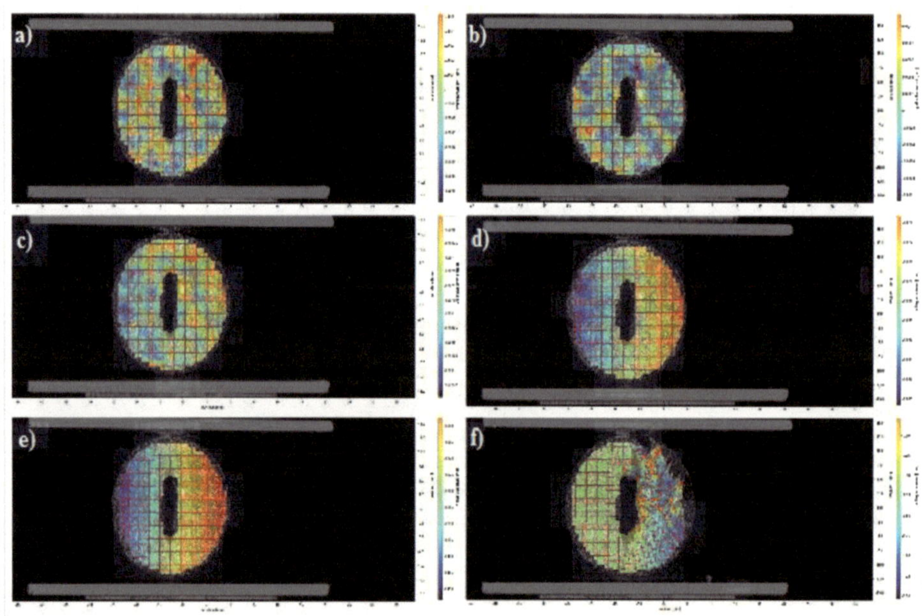

Figure 7.
The image of correlation.

that the angle 22°C verifies the pure mode II. The value of this angle is in the interval of 20° and 30°. A very good agreement is shown between the theoretical predictions and the experimental results. Applying the load, the same crack propagation is carried out in the numerical and experimental element.

3.2 Digital image correlation

For mode I, a series of images is taken with a CCD camera and digitalized and then compared to the reference image. It has been verified that in Mode I loading, the crack surfaces separate symmetrically and the crack front propagates in the direction of the crack plane (**Figure 7**).

3.3 Determination of stress intensity factor under mode I

Since the CSTBD specimen is closely linked to fracture toughness and stress intensity factor, this sub section is there for compare different solutions of stress intensity factor under mode I. We used the experimental data of the CSTBD samples to define KI, by applying the previous formulas Eq. (3) using two different term of Y (6) and Eq. (7) using the different term of dimensionless stress intensity factor (NI) (9) and (10), for different percentages of TCP additive calculation is launched for crack length a/R = 0.4.

According to our previously work [9], the used composite specimens reached their optimum in mechanical properties at 1600° C. **Figure 8** illustrates the evolution of the stress intensity factor under mode I fracture in relation to the percentage of TCP under optimal conditions at 1600°C for 1 hour using different methods. We note that the Cherepanov, SIF values are close to those by handbook, Shetty and Fowel and al. Hence, K_I (Cherepanov) is basically consistent with K_I (handbook) and K_I (Shetty et all) and shows a good compromise in the results.

For the study of the effect of TCP, this figure has illustrated that the stress intensity factor K_I increases with the addition of 10 wt.% of TCP until 8.452 MPa $m^{1/2}$ using the formula mentioned in Cherepanov's book. Beyond this percentage of TCP, the overall stiffness falls gradually.

The initiation and propagation of each crack depends on the type of solicitation. According to this test condition, cracks propagates in a parallel manner to the direction of the notch, and as soon as it intersects with the surface, the sample is divided into two parts. These cracks are generated by principal stresses, under mode I loading (As shown in **Figure 9**).

3.4 Determination of stress intensity factor under mode II

In this research, the loading angle corresponding to mode II deformation in the CSTBD specimens is approximately equal to 22 (deg.) for α = 0.4. Consequently, the mode II fracture tests were performed according to these loading angle and then K_{II} is determined for pure mode II. In this test condition, a crack propagates according

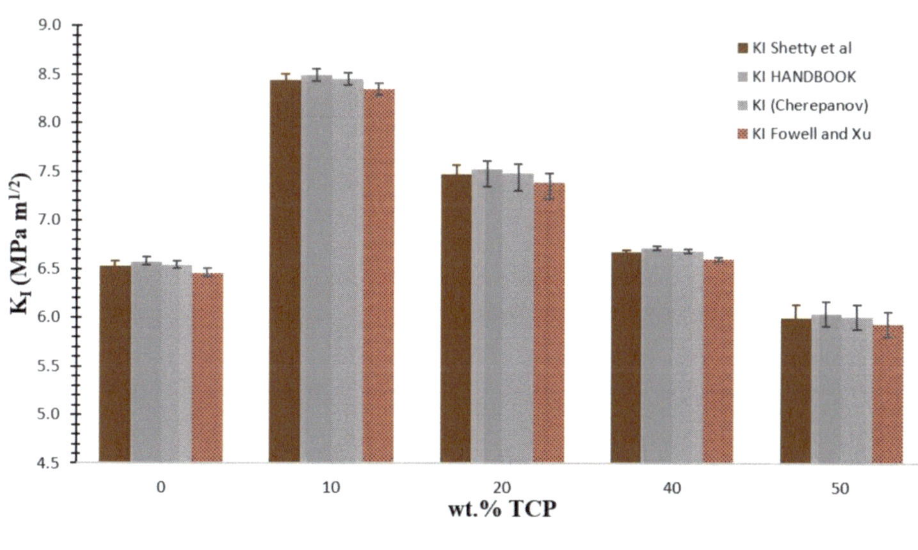

Figure 8.
Mode I stress intensity factor versus percentage of TCP additive in Al2O3 using different methods.

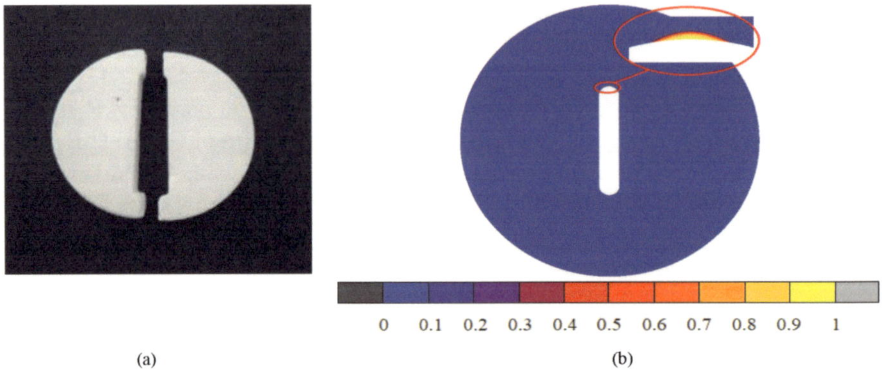

Figure 9.
Influence of mode I (opening mode) on crack initiation and propagation after the fracture tests: (a) experimental fracture and (b) numerical fracture.

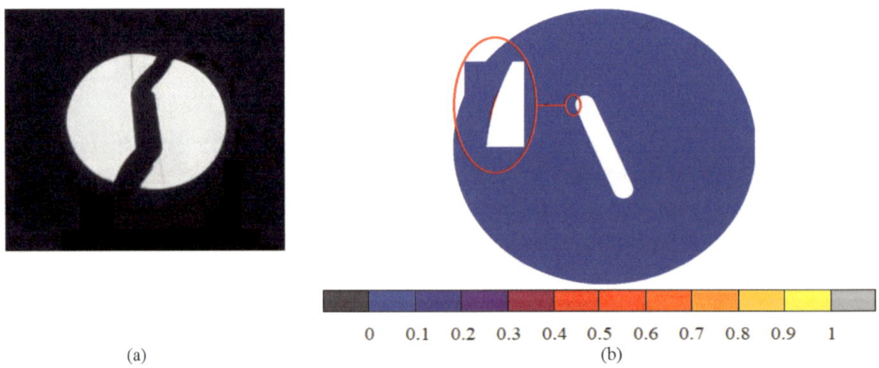

Figure 10.
Influence of mode II (shearing mode) on crack initiation and propagation after the fracture tests: (a) experimental fracture and (b) numerical detection by successive simulations of the mode II loading angle resulting in pure shearing mode effects.

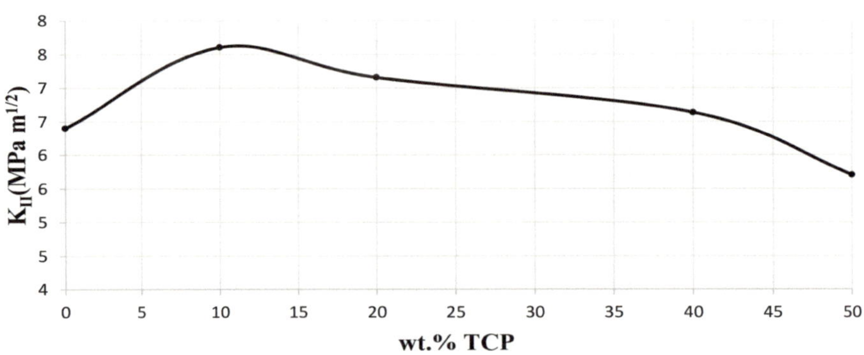

Figure 11.
Mode II stress intensity factor versus percentage of TCP additive (in Al_2O_3).

to mode II test and the same crack propagation is carried out in the numerical and experimental part. (see **Figure 10**).

Analytical analysis for this geometry is accomplished by using Eq. (11). **Figure 11** presents the test results for the calculation of the stress intensity factor for the same crack length with different percentages of TCP respectively. The stress intensity factor mode II fracture values of Al_2O_3-wt.% TCP composites are comprised between 5.8 MPa $m^{1/2}$ and 7.6 MPa $m^{1/2}$. The lowest toughness is obtained with the 50 wt.% TCP, while the highest one is approached with the 10 wt.% TCP (7.6 MPa $m^{1/2}$) (see **Figure 11**).

3.5 Determination of the fracture toughness K_{IC}

For the SCB test, after the crack starts from the disc center at the maximum load, the crack propagates symmetrically ahead the loading diameter. K_{IC} is determined by applying Eq. (1), the K_{IC} is calculated for different percentages of TCP.

Figure 12 shows the fracture toughness of different amounts of TCP added to Alumina. As sintered at 1600°C/1 h, The best value of the fracture toughness value of Al_2O_3/10 wt.% TCP composites is 8.76 MPa $m^{1/2}$.

3.6 S.E.M characterization of sintered samples

The microstructure of the sintered specimens was observed by scanning electronic microscopy (SEM). **Figure 13** shows the S.E.M micrographs of the Alumina (Al_2O_3) reinforced with the Tricalcium phosphate (TCP) with different additives amounts sintered at 1600°C for 1 hour. This technique helps to investigate the porosity and the texture of any biomaterial. These micrographs show the coalescence between alumina grains produced with all the percentages of added TCP (**Figure 13(a–f)**).

These micrographs show the difference between fracture surface of Alumina, Alumina −10 wt.% TCP and Alumina −50 wt.% TCP samples (**Figure 13(a–f)**). The SEM micrographs of the pure Alumina sintered without TCP show an intergranular porosity (**Figure 13a** and **b**). The significant improvement of the characteristics of the Al_2O_3–10 wt.% TCP samples can be explaining by a coalescence between Alumina grains produced with all the percentages of added TCP. In addition, the formed spherical pores demonstrate that a liquid phase was formed a 1600°C (**Figure 13c** and **d**). For this composition, one notices also the absence of micro-cracking and the reduction of the sizes of the pores that achieves higher

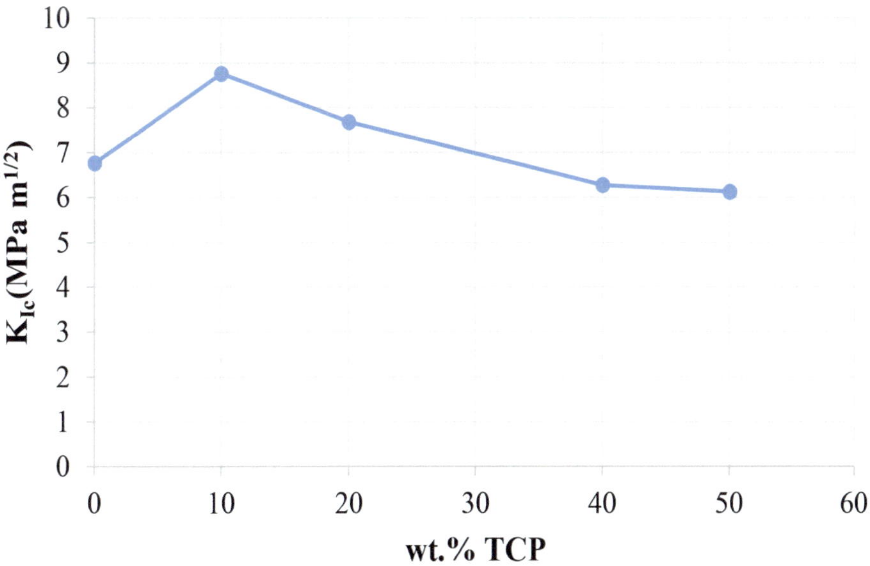

Figure 12.
Fracture toughness (K_{IC}) as function of the percentage of TCP.

Figure 13.
SEM micrographs of different bioceramic composites sintered at 1600°C for 1 h: (a) and (b) Al_2O_3, (c) and (d) Al_2O_3–10 wt.% TCP, (e) and (f) Al_2O_3–50 wt.% TCP.

densities and decrease the grain size. For this reason, the Alumina-10 wt.% TCP composite presents excellent mechanical properties.

The strength started to decrease with higher percentage of TCP (20, 40, and 50 wt.%). A particular relation between grain size and mechanical strength in sintered alumina-TCP composites was found.

4. Conclusion

This work aimed at studying the fracture behavior of the Alumina-Tricalcium phosphate. From the main results, the following conclusions can be drawn:

The fracture toughness KIC of the samples contains a crack were assessed using Semi Circular Bending tests. The best fracture properties in terms of fracture toughness, were obtained for Al2O3–10 wt.% TCP composition. The best value is 8.76 MPa $m^{1/2}$.

The CSTBD specimen is an appropriate test specimen for fracture tests of U-shaped notches particularly when the notch is subjected to pure mode II loading.

The CSTBD specimen was employed for mixed Mode I and II fracture studies. The numerical analysis of Alumina-TCP subjected to static indentation by a spherical indenter was presented. From a CDM based constitutive modeling, the anisotropic damage mechanisms developed in the specimens were examined through the principal (mode I) and shear stresses (mode II). The stress intensity factor under mode I and II of the Alumina-TCP composites increase with the amount of TCP until 10 wt.%.

The initiation and propagation of each crack depends on the type of solicitation. The geometry model used includes a Brazilian disc containing a crack with two different orientations: 0° for tensile mode (mode I), 22° for shear mode (mode II). The predicted directions were found in good agreement with the experimental observations reported in the literature.

Author details

Barkallah Rachida[1]*, Rym Taktak[1], Noamen Guermazi[2], Fahmi Zaïri[3] and Jamel Bouaziz[1]

1 Laboratoire des MatériauxAvancés (LMA), EcoleNationaled'Ingénieurs de Sfax (ENIS), Université de Sfax, Tunisia

2 Laboratoire de Génie des Matériaux et Environnement (LGME), Ecole Nationale d'Ingénieurs de Sfax (ENIS), Université de Sfax, Tunisia

3 Lille University, Civil Engineering and Geo-Environmental Laboratory (ULR 4515 LGCgE), Lille, France

*Address all correspondence to: barkallah_rachida1@live.fr

References

[1] Zeineddine Hussein A, Frush Todd J, Saleh Zeina M, et al. Applications of tissue engineering in joint arthroplasty: current concepts update. Orthopedic Clinics. 2017; 48(3): 275-288. doi: 10.1016/j.ocl.2017.03.002.

[2] Shin YS, Choi JW, Park JK, et al. Tissue-engineered tracheal reconstruction using mesenchymal stem cells seeded on a porcine cartilage powder scaffold. Ann Biomed Eng. 2015;43(4):1003–13. DOI: 10.1007/s10439-014-1126-1

[3] Sun H, Liu W, Zhou G, et al. Tissue engineering of cartilage, tendon and bone. Frontiers of medicine. 2011;5(1): 61–9.https://doi.org/10.1007/s11684-011-0122-1

[4] Ayatollahi MR, Aliha MRM, Saghafi H. An improved semicircular bend specimen for investigating mixed mode brittle fracture. EngFract Mech. 2011;78(1):110–123. doi.org/10.1016/j.engfracmech.2010.10.001 11.

[5] XIE, Yousheng, CAO, Ping, JIN, Jin, et al. Mixed mode fracture analysis of semi-circular bend (SCB) specimen: a numerical study based on extended finite element method. Computers and Geotechnics. 2017;82:157–172.https://doi.org/10.1016/j.compgeo.2016.10.012

[6] AYADI Ibticem, AYED Foued Ben. Sintering and the mechanical properties of the tricalcium phosphate–titania composites. journal of the mechanical behavior of biomedical materials. 2015; 49:129-140.DOI: 10.1016/j.jmbbm.2015.05.001

[7] R D AGaasbeek, H G Toonen, RJ Van Heerwaarden, P Buma, Mechanism of bone incorporation of β-TCP bone substitute in open wedge tibial osteotomy in patients. Biomaterials. 2005; 26:6713–6719.DOI: 10.1016/j.biomaterials.2005.04.056

[8] J.L. Runyan, S.J. Bennison, Fabrication of flaw-tolerant aluminiumtitanate-reinforced alumina, J. Eur. Ceram. Soc. 7 (1991) 93–99.h ttps://doi.org/10.1016/0955-2219(91) 90006-L

[9] RBarkallah, RTaktak, NGuermazi, FZaïri, JBouaziz, FZaïri. Manufacturing and mechanical characterization of Al_2O_3/β-TCP/TiO_2 biocomposite as a potential bone substitute. Int J AdvManuf Technol. 2017; 3369–3380. DOI:10.1007/s00170-017-1434-3

[10] YAO Wei, XU Ying, YUChangyiet al. A dynamic punch-through shear method for determining dynamic Mode II fracture toughness of rocks. Engineering Fracture Mechanics. 2017; 176: 161-177. https://doi.org/10.1016/j.engfracmech.2017.03.012

[11] AyatollahiM R, Torabi A R. Determination of mode II fracture toughness for U-shaped notches using Brazilian disc specimen. International Journal of Solids and Structures. 2010; vol. 47, no 3-4, p. 454-465. https://doi.org/10.1016/j.ijsolstr.2009.10.012

[12] QZ Wang, Formula for calculating the critical stress intensity factor in rock fracture toughness tests using cracked chevron notched Brazilian disc (CCNBD) specimens, Int JRock Mech Min Sci. 2010; 47:1006–1011. DOI: 10.1016/j.ijrmms.2010.05.005

[13] SH Chang, CLee, S Jeon, Measurement of rock fracture toughness under modes I and II and mixed-mode conditions by using disc-type specimens. Eng. Geol. 2002; 66:79–97. DOI:10.1016/S0013-7952(02)00033-9

[14] Ghavidel NA, Memarian H, Mohamadi S, Heydarizadeh M. Analytical Solution for Stress Field and Intensity Factor in CSTBD under Mixed Mode Conditions, Int. J. Min. & Geo-

Eng.2014; 48(1):55-68. Doi:10.22059/IJMGE.2014.51806i

[15] Ismail Jewan, ZaïriFahmi, Naït-abdelazizMoussa et al. Computational modelling of static indentation-induced damage in glass. ComputationalMaterials Science. 2008; 42(3): 407-415.DOI:10.1016/j.commatsci.2007.08.006

[16] Belrhiti Y, Dupre JC, Pop O, et al. Combination of Brazilian test and digital image correlation for mechanical characterization of refractory materials. Journal of the European Ceramic Society. 2017; 37(5):2285-2293. DOI: 10.1016/j.jeurceramsoc.2016.12.032

[17] J C Dupré, PDoumalin, H AHusseini, A Germaneau, FBrémand, Displacement discontinuity or complex shape of sample: assessment of accuracy and adaptation of local DIC approach. Strain. 2015; 51:391–404. DOI:10.1111/str.12150

[18] Sakka S, Ben Ayed F, Bouaziz J. Mechanical properties of tricalcium phosphate-alumina composites. Mater Sci Eng.2012; 28: 012028DOI: 10.1088/1757-899X/28/1/012028

[19] Kuruppu Mahinda D et CHONG Ken P. Fracture toughness testing of brittle materials using semi-circular bend (SCB) specimen. Engineering Fracture Mechanics. 2012; 91:133-150. DOI:10.1016/j.engfracmech.2012.01.013

[20] S H Chang, C Lee, S Jeon, Measurement of rock fracture toughness under modes I and II and mixed-mode conditions by using disc-type specimens, EngGeol. 2002; 66:79–97. DOI:10.1016/S0013-7952(02)00033-9

[21] FowellRJ, Chen, JF. The third chevron-notch rock fracture specimen the cracked chevron notched Brazilian disk. Proc. 31st U.S. Symp. Rock. Balkema. Rotterdam. 1990; 295–302. DOI:10.1016/j.ijrmms.2011.09.019

[22] FowellRJ, XUC. The cracked chevron notched Brazilian disc test geometrical considerations for practical rock fracture toughness measurement. Int. J. Rock.Mech. Mineral Sci. Geomech. Abstr. 1993; 30(7):821 – 824.https://doi.org/10.1016/0148-9062(93)90029-D

[23] Wang QZ, Jia XM, Kou SQ, Zhang ZX, Lindqvist P A. More accurate stress intensity factor derived byfinite element analysis for the ISRM suggested rock fracture toughness specimenCCNBD, Int. J. Rock. Mech. Min.Sci, 2003; 40:233–241. DOI: 10.1016/S1365-1609(02)00131-4

[24] Iqbal MJ, Mohanty B, Experimental calibration of stress intensity factors of the ISRM suggested cracked chevron-notched Brazilian disc specimen used for determination ofmode-I fracture toughness, Int J Rock Mech Min Sci, 2006;43:1270–1276.DOI: 10.1016/j.ijrmms.2006.04.014

[25] Cherepanov GP. Mechanics of brittle fracture. New York: McGraw-Hill; 1979. https://doi.org/10.1007/BF00037929

[26] Tada H, Paris PS, Irwin GR. The stress analysis of cracks handbook, 2nd ed. St Louis: Paris Productions. 1985.

[27] Eftekhari M, Bahbanan A, Hashemolhosseini H. Extended finite element simulation of crack propagation in cracked Brazilian disc, J Min Env. 2015;6(1):95-102.Doi: 10.22044/JME.2015.365

[28] Shetty DK, RosenfiledAR, Duckworth, W H, Fracture toughness of ceramics measured by a chevron notch diametral compression test, J. Am. Ceram. Soc, 1985;68 (12):C325 – C327.https://doi.org/10.1111/j.1151-2916.1985.tb10135.x

[29] AtkinsonC, SmelserR E, Sanchez J. Combined mode fracture via the

cracked Brazilian disc test, Int JFract 1982;18):279–91.https://doi.org/10.1007/BF00015688

[30] P Bing, XHui-min, H Tao, A. Asundi, Measurement of coefficient of thermalexpansion of films using digital image correlation method, Polym. Test. 2009;28:75–83. DOI:10.1016/j.polymertesting.2008.11.004

[31] YBelrhiti, A Germaneau, P Doumalin, J CDupré, O Pop, M Huger, TChotard.Characterization of the mechanical behavior of magnesia spinel refractoriesusing image correlation, Proc. of Int. Conf. UNITECR.2013;13:6–41. DOI:10.1111/ijac.12307

[32] L MKachanov, Time of the rupture process under creep conditions, IzvestiyaAkademiiNauk USSR.1958;8: 26–31 (in Russian).

[33] Lemaitre Jean et ChabocheJean-Louis. Mecanique des materiaux solides. Dunod. Mechanics of Solid Materials, Springer, 1985.

[34] Hamza Samir. Etude du comportement en fatigue en compression des biocéramiques (Al_2O_3, ZrO_2) utilisées pour la conception des prothèses ostéo-articulaires. Thèse de doctorat. Université Paul Verlaine-Metz. 2002.https://hal.univ-lorraine.fr/tel-01749598

Chapter 5

XEFGM Fracture Analysis of Functionally Graded Materials under Mixed Mode and Asymmetric Loading

Nathera A. Saleh and Haider Khazal

Abstract

This research describes crack analysis in the functionally graded materials (FGMs) by adopting the extended element free Galerkin method (XEFGM) under mixed mode and asymmetric loading. These loads are somewhat similar to fatigue loads because, numerically, they are load values from zero and then directly to the critical load. The meshfree method can be easily simulated the fracture problems against the traditional numerical method because it is not dependent on mesh. Triangles technique in the process of numerical integration at regions of discontinuity, functions of enrichment, and as well as the appropriate support field to contain numerical points and nodes to from the shape functions are used in this study. In addition, incompatible interaction integration technique has used to determine the stress intensity factors (SIFs). Two study cases with different crack positions were studied and compared with the experimental works of the relevant reference literature, where accurate and identical results were obtained.

Keywords: Crack propagation, incompatible interaction integration technique, meshfree method, functionally graded materials

1. Introduction

Functionally graded materials (FGMs) arose from the realization that it was necessary to meet ultra-high temperature and cryogenic requirements. The goals of strength, flexibility, and fatigue resistance inspired the early research. This aim in achieving a smooth and perfect spatial variation is most effectively met by gathering different materials with those favorable characteristics, thereby avoiding the detriment effect such as stress concentration and residual stress found in discrete interface. The gradual change in material properties from the original composite has been shown to increase efficiency by mitigating failure and maintaining the intended benefits of merging two or more different materials. Functionally graded materials are found in nature, for example in bones, teeth, wood, and bamboo [1]. As a result of the microstructure and mechanical properties of functionally graded materials (FGMs) that different from their corresponding from the conventional composites, they have to be used in many engineering, military, medical and space sciences applications. Numerical methods have been used and applied in the analysis of fracture problems in these materials because they have the ability to give realistic results [2–11]. Popovich et al. [12], Zhao et al. [13] studied fatigue load that exerted on certain types of functionally graded materials. Most recently, digital image correlation technique was used with numerical verification on a stepwise functionally graded material made of glass and epoxy to find the path growth of the crack and the stress concentration values [14, 15]. On the other hand, the optical method was used to analyze the crack path in a material made continuously from the graded materials and the stress concentration factor and T-stress were also calculated [16]. One of these methods that have been adopted in this research is the extended weak Galerkin formulation-element free method (XEFGM). The element-free Galerkin method with the sub-triangle technique to enhance the accuracy of the Gauss squared near the crack to determine stress intensity factors in isotropic and anisotropic materials can give accurate results in analyzing fracture structures. The incompatible interaction integral technique is applied to extract stress intensity factors (SIFs). The method of sub-triangulation is used to hence the discontinuity terms. With low degrees of freedom, results can be obtained that are highly consistent with those of the relevant reference literature.

2. Field Equations

In elastic materials, Hooke's law can be employed in the following equation that clears the relationship between the stresses and strains in specific material.

$$\boldsymbol{\sigma} = \mathbf{D}\boldsymbol{\varepsilon} \tag{1}$$

$\boldsymbol{\sigma}$ and $\boldsymbol{\varepsilon}$ are the stress and strain respectively; while $\mathbf{D}$ represents the material matrix that varied with the displacement x in the functionally graded materials can be find as:

$$\mathbf{D} = \frac{E(x)}{1-\nu^2}\begin{bmatrix} 1 & \nu(x) & 0 \\ \nu & 1 & 0 \\ 0 & 0 & (1-\nu(x))/2 \end{bmatrix} \quad \text{(Plane stress)} \tag{2}$$

and

$$\mathbf{D} = \frac{E(x)(1-\nu(x))}{(1+\nu(x))(1-2\nu(x))}\begin{bmatrix} 1 & \frac{\nu(x)}{1-\nu(x)} & 0 \\ \frac{\nu(x)}{1-\nu(x)} & 1 & 0 \\ 0 & 0 & \frac{1-2\nu(x)}{2(1-\nu(x))} \end{bmatrix} \quad \text{(Plane strain)} \tag{3}$$

Where E and ν represent Young's Modulus and Poisson's ratio that varied with the displacement x respectively, Eqs. (2) and (3) are useful because they give a change in the properties of the material at each point of the material in relation to the displacement.

Eq. (1) can be put as:

$$\boldsymbol{\varepsilon} = \mathbf{C}\boldsymbol{\sigma} = \begin{bmatrix} C_{11} & C_{12} & C_{16} \\ C_{12} & C_{22} & C_{26} \\ C_{16} & C_{26} & C_{66} \end{bmatrix} \begin{Bmatrix} \sigma_{xx} \\ \sigma_{yy} \\ \sigma_{xy} \end{Bmatrix} \tag{4}$$

where **C** represents the compliance matrix.

In FGM

$$E = E(x_1, x_2) = E(x) \quad (5)$$

$$\upsilon = \upsilon(x_1, x_2) = \upsilon(x) \quad (6)$$

In the above two equations, it is clear that the Young's Modulus and Poisson's ratio change with the displacement and this gives the impression for the behave of properties of the functionally graded materials as shown in **Figure 1**.

The stresses and displacements in functionally graded materials near the crack tip can be obtained by depending on angular stress and displacement functions as [2]:

$$\sigma_{11} = \frac{1}{\sqrt{2\pi r}}\left[K_I f_{11}^I(\theta) + K_I f_{11}^{II}(\theta)\right] \quad (7)$$

$$\sigma_{22} = \frac{1}{\sqrt{2\pi r}}\left[K_I f_{22}^I(\theta) + K_{II} f_{22}^{II}(\theta)\right] \quad (8)$$

$$\sigma_{12} = \frac{1}{\sqrt{2\pi r}}\left[K_I f_{12}^I(\theta) + K_{II} f_{12}^{II}(\theta)\right] \quad (9)$$

$$u_1 = \frac{1}{G_{tip}}\sqrt{\frac{r}{2\pi}}\left[K_I g_1^I(\theta) + K_{II} g_1^{II}(\theta)\right] \quad (10)$$

$$u_2 = \frac{1}{G_{tip}}\sqrt{\frac{r}{2\pi}}\left[K_I g_2^I(\theta) + K_{II} g_2^{II}(\theta)\right] \quad (11)$$

$$G_{tip} = \frac{E_{tip}}{\left[2\left(1 + \nu_{tip}\right)\right]} \quad (12)$$

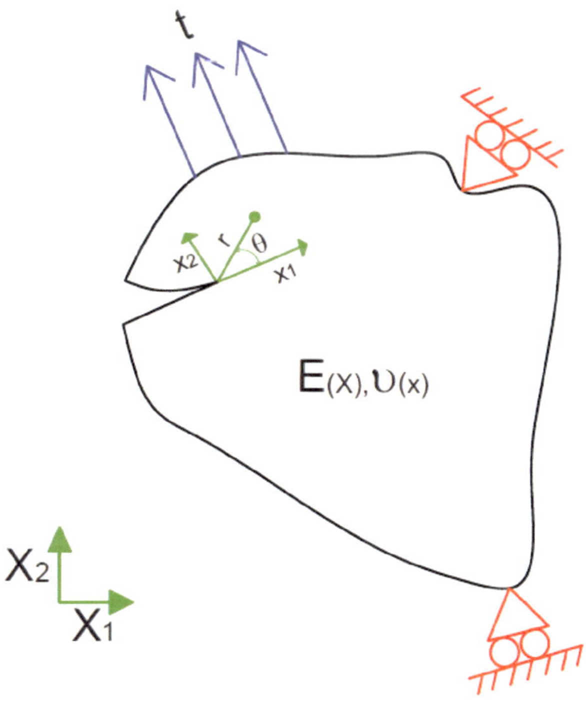

Figure 1.
FGM body with crack,

The mechanical properties in Eq. (11) to (13) will be extracted at the tip of crack. $f^{I}_{ij}(\theta), f^{II}_{ij}(\theta), g^{I}_{ij}(\theta)$ and $g^{II}_{ij}(\theta)(i, j = 1, 2)$ represent the standard angular functions [17, 18]. Eqs. (7) to (13) can calculate the values of stresses and displacements at each point of the material supported by the enrichment functions.

3. XEFGM Structure

In the two-dimensional solid elasticodynamics as depicted in **Figure 2**, the equilibrium equation that gives the relationship between the stresses and traction load $\boldsymbol{t}$ can be depicted in matrix form as:

$$\boldsymbol{L}^T\boldsymbol{\sigma} + \boldsymbol{b} = 0 \quad in\ problemd\ dmain\ \Omega \tag{13}$$

Where the following Eqs. (15) to (17) represent boundary equations.

$$\boldsymbol{\sigma n} = \bar{\boldsymbol{t}} \qquad on \quad \Gamma_t \tag{14}$$

$$\boldsymbol{u} = \bar{\boldsymbol{u}} \qquad on \quad \Gamma_u \tag{15}$$

$$\boldsymbol{\sigma n} = 0 \qquad on \quad \Gamma_e \tag{16}$$

And the differential operator matrix $\boldsymbol{L}$ is

$$L = \begin{bmatrix} \frac{\partial}{\partial x} & 0 \\ 0 & \frac{\partial}{\partial y} \\ \frac{\partial}{\partial y} & \frac{\partial}{\partial x} \end{bmatrix} \tag{17}$$

EFGM adopted the moving least squares (MLS) approximation [19] to find shape function of the numerical method. The governing equation of the relevant problem:

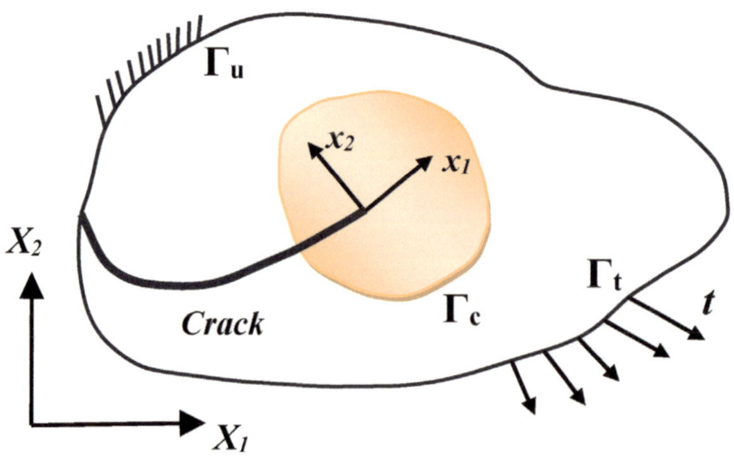

Figure 2.
2D cracked body.

$$\int_{\Omega} (\boldsymbol{L}\delta\boldsymbol{u})^T(\boldsymbol{DLu})d\Omega - \int_{\Omega} \delta\boldsymbol{u}^T\boldsymbol{b}d\Omega - \int_{\Gamma_t} \delta\boldsymbol{u}^T\bar{\boldsymbol{t}}d\Gamma - \int_{\Gamma_u} \delta\lambda^T(u - \bar{u})d\Gamma - \int_{\Gamma_u} \delta\boldsymbol{u}^T\lambda d\Gamma = 0 \tag{18}$$

where λ represents the Lagrange multiplier variable. The MLS shape function ϕ_i at node i a point $\boldsymbol{x}$ can be defined as [20]:

$$\phi_i(\boldsymbol{x}) = \boldsymbol{p}^T(\boldsymbol{x})[\boldsymbol{A}(\boldsymbol{x})]^{-1}w(\boldsymbol{x} - \boldsymbol{x}_i)\boldsymbol{p}(\boldsymbol{x}_i) \tag{19}$$

$\boldsymbol{p}(\boldsymbol{x})$ is the basis function

$$\boldsymbol{p}^T(\boldsymbol{x}) = [1 \;\; xy] \tag{20}$$

And $\boldsymbol{A}$ can be extracted as

$$\boldsymbol{A}(\boldsymbol{x}) = \sum_{i=1}^{n} w(\boldsymbol{x} - \boldsymbol{x}_i)\boldsymbol{p}(\boldsymbol{x}_i)\boldsymbol{p}^T(\boldsymbol{x}_i) \tag{21}$$

And

$$w(r) = \begin{cases} \frac{2}{3} - 4r_s^2 + 4r_s^3 & r_s \le \frac{1}{2} \\ \frac{4}{3} - 4r_s + 4r_s^2 - \frac{4}{3}r_s^3 & \frac{1}{2} < r_s \le 1 \\ 0 & r_s > 1 \end{cases} \tag{22}$$

Where r_s is the size of the influence domain for node i.

The externally enriched displacement approximation $\boldsymbol{u}^h$ of a model point x [21–23]:

$$\boldsymbol{u}^h(\mathbf{x}) = \sum_{i=1}^{n} \phi_i(\mathbf{x})\boldsymbol{u}_i + \sum_{k=1}^{m_t} \phi_k \sum_{\alpha=1}^{4} Q_\alpha(\mathbf{x})\boldsymbol{b}_k \tag{23}$$

For isotropic FGMs, enrichment functions are applied to enrich the MLS formulation [24, 25]:

$$Q(r,\theta) = \left(\sqrt{r}\cos\left(\frac{\theta}{2}\right), \sqrt{r}\sin\left(\frac{\theta}{2}\right), \sqrt{r}\sin\left(\frac{\theta}{2}\right)\sin(\theta), \sqrt{r}\cos\left(\frac{\theta}{2}\right)\sin(\theta)\right) \tag{24}$$

Therefore the final discretized system equations,

$$\begin{bmatrix} \boldsymbol{K} & \boldsymbol{Q} \\ \boldsymbol{Q}^T & \boldsymbol{0} \end{bmatrix} \begin{bmatrix} \boldsymbol{U} \\ \lambda \end{bmatrix} = \begin{bmatrix} \boldsymbol{F} \\ \boldsymbol{q} \end{bmatrix} \tag{25}$$

The vectors in Eq. (26)

$$\boldsymbol{Q} = -\int_{\Gamma_u} \boldsymbol{N}^T\phi d\Gamma \tag{26}$$

$$\boldsymbol{q} = -\int_{\Gamma_u} \boldsymbol{N}^T \overline{\boldsymbol{u}} \tag{27}$$

$$\lambda(\mathbf{x}) = \sum_{i=1}^{n_\lambda} N_i(\mathbf{x})\lambda_i \tag{28}$$

U is the global displacement vector can be defined as:

$$U = \{\boldsymbol{u}\ \boldsymbol{b}_1\ \boldsymbol{b}_2\ \boldsymbol{b}_3\ \boldsymbol{b}_4\}^T \tag{29}$$

Where b_1 to b_4 represent the enrichment, function terms.
And

$$\boldsymbol{K}_{ij}^n = \begin{bmatrix} K_{ij}^{uu} & K_{ij}^{ub} \\ K_{ij}^{bu} & K_{ij}^{bb} \end{bmatrix} \tag{30}$$

$$\boldsymbol{F}_i^n = \left\{ \boldsymbol{F}_i^u\ \ \boldsymbol{F}_i^{b_1}\ \ \boldsymbol{F}_i^{b_2}\ \ \boldsymbol{F}_i^{b_3}\ \ \boldsymbol{F}_i^{b_4} \right\}^T \tag{31}$$

where

$$\boldsymbol{K}_{ij}^{rs} = \int_{\Omega} (\boldsymbol{B}_i^r)^T \boldsymbol{D}\boldsymbol{B}_j^s d\Omega \quad (r,s = u,b) \tag{32}$$

$$\boldsymbol{F}_i^u = \int_{\Omega} \phi_i^t \boldsymbol{b} d\Omega + \int_{\Gamma_t} \phi_i^T \bar{\boldsymbol{t}} d\Gamma \tag{33}$$

$$\boldsymbol{F}_i^{b_a} = \int_{\Omega} \phi_i^T Q_a \boldsymbol{b} d\Omega + \int_{\Gamma_t} \phi_i^T Q_a \bar{\boldsymbol{t}} d\Gamma \ \ (a = 1,2,3,4) \tag{34}$$

$\boldsymbol{B}_i^u$ and $\boldsymbol{B}_i^b$ are matrices of shape function derivatives:

$$\boldsymbol{B}_i^u = \begin{bmatrix} \phi_{i,x} & 0 \\ 0 & \phi_{i,y} \\ \phi_{i,y} & \phi_{i,x} \end{bmatrix} \tag{35}$$

$$\boldsymbol{B}_i^b = \begin{bmatrix} \boldsymbol{B}_i^{b_1} & \boldsymbol{B}_i^{b_2} & \boldsymbol{B}_i^{b_3} & \boldsymbol{B}_i^{b_4} \end{bmatrix} \tag{36}$$

$$\boldsymbol{B}_i^u = \begin{bmatrix} (\phi_i Q_\alpha)_{,x} & 0 \\ 0 & (\phi_i Q_\alpha)_{,y} \\ (\phi_i Q_\alpha)_{,y} & (\phi_i Q_\alpha)_{,x} \end{bmatrix} \tag{37}$$

$$\alpha = 1,2,3,4 \tag{38}$$

$$\boldsymbol{\varepsilon} = \boldsymbol{L}\mathbf{u}^{\mathrm{h}} \tag{39}$$

$$\boldsymbol{\sigma} = \boldsymbol{D\varepsilon} \tag{40}$$

Eq. (40) and (41) represent stress and strain for whole body of FG material calculated by numerical method.

4. Fracture Criteria

Incompatibility formulation is applied to extracted J-integral [26, 27].

$$\sigma_{ij} = d_{ijkl}(x)\varepsilon_{kl} \ , \qquad \varepsilon_{ij} \neq \frac{1}{2}\left(u_{i,j} + u_{\ j,i}\right) \ , \ \sigma_{ij,j} = 0 \tag{41}$$

where

$$\varepsilon_{ij} = C_{ijkl}(x)\sigma_{kl} \quad (i,j,k,l = 1,2,3) \tag{42}$$

with

$$\mathbf{C} = \begin{bmatrix} C_{11} & C_{12} & C_{16} \\ C_{12} & C_{22} & C_{26} \\ C_{16} & C_{26} & C_{66} \end{bmatrix} = \begin{bmatrix} C_{1111} & C_{1122} & 2C_{1112} \\ C_{2211} & C_{2222} & 2C_{2212} \\ 2C_{1211} & 2C_{1222} & 4C_{1212} \end{bmatrix} \tag{43}$$

From **Figure 3**:

$$J = \int_A \left(\sigma_{ij}u_{i,1} - w\delta_{1j}\right)q,_{j}dA + \int_A \left(\sigma_{ij}u_{i,1} - w\delta_{1j}\right)_{,j}qdA \tag{44}$$

w is the strain energy density:

$$w = \frac{1}{2}\sigma_{ij}\varepsilon_{ij} \tag{45}$$

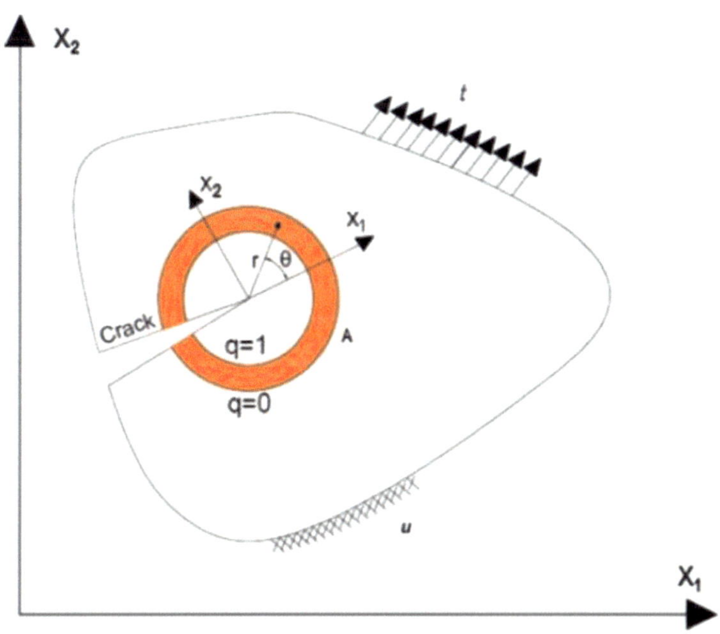

Figure 3.
An integral contour at the tip of the crack.

The interaction integral and J integral can be defined [28–30]:

$$M = \int_A \left\{ \sigma_{ij} u_{i,1}^{aux} + \sigma_{ij}^{aux} u_{i,1} - \frac{1}{2}\left(\sigma_{ik} u_{ik}^{aux} + \sigma_{ik}^{aux} u_i\right)\delta_{1j} \right\} q,_j dA + \int_A \left\{ \sigma_{ij}\left(c_{ijkl}^{tip} - c_{ijkl}(x)\right)\sigma_{kl,1}^{aux} \right\} q dA \quad (46)$$

$$J_{local} = \left(K_I^2 + K_{II}^2\right)/E_{tip} \quad (47)$$

$$J_{local}^s = \frac{\left(K_I + K_I^{aux}\right)^2 + \left(K_{II} + K_{II}^{aux}\right)^2}{E_{tip}} = J_{local} + J_{local}^{aux} + M_{local} \quad (48)$$

$$J_{local}^{aux} = \left[\left(K_I^{aux}\right)^2 + \left(K_{II}^{aux}\right)^2\right]/E_{tip} \quad (49)$$

and M_{local} is given by

$$M_{local} = 2\left(K_I K_I^{aux} + K_{II} K_{II}^{aux}\right)/E_{tip} \quad (50)$$

$$\begin{aligned} K_I &= M_{local}^{(1)} E_{tip}/2, \qquad \left(K_I^{aux} = 1.0, K_{II}^{aux} = 0.0\right), K_{II} \\ &= M_{local}^{(2)} E_{tip}/2, \qquad \left(K_I^{aux} = 0.0, K_{II}^{aux} = 1.0\right). \end{aligned} \quad (51)$$

The Eq. (51) is used to calculate the stress intensity factors during the fracture analysis in functionally graded materials. The crack propagation criterion (Maximum hoop stress) was applied by the depend on procedure that was adopted by Erdogan and Sih [31].

5. Numerical examples

5.1 Example 1

The first example is that there is a crack in the FGM beam that undergoes a three-point bend as shown in **Figure 4**, and in this problem, the beams are homogeneous and gradient (along with the X_2 direction). **Figure 4a** and **b** show the geometry of the sample and BCs for two different boundary conditions: the states (a) and (b), respectively. Also, this figure shows the complete nodal distributions, and the adaptive background cells visualize the distribution of Gauss points and fertilization nodes around the crack for Case (a). Note that the nodal and background cell distributions are valid for both conditions. The material properties (**Table 1**) of the monolithic beam used are as follows:

$$E = 2890\,MPa, \nu = 0.4, K_{Ic} = 1.09\,MPa\sqrt{m}.$$

A 64 x 28 back grid and 1856 non-uniform distribution nodes are adopted in this case (**Figure 5**). 2875 nodes in finite element method was previously used by Kim and Paulino [3]. **Figure 6** depicts the comparison of the crack path of a homogeneous case (b) beam obtained by current work with the experimental results reported by Galvez et al. [32] With numerical simulation by [3]. The reasonably well output between the numerical and experimental results are obtained. Note that in this case, the gradient of the material does not affect the path of the crack. **Figure 7** shows the effect of increasing the slit length by current numerical simulations of (b) condition on the slit path compared to the experiment available for a

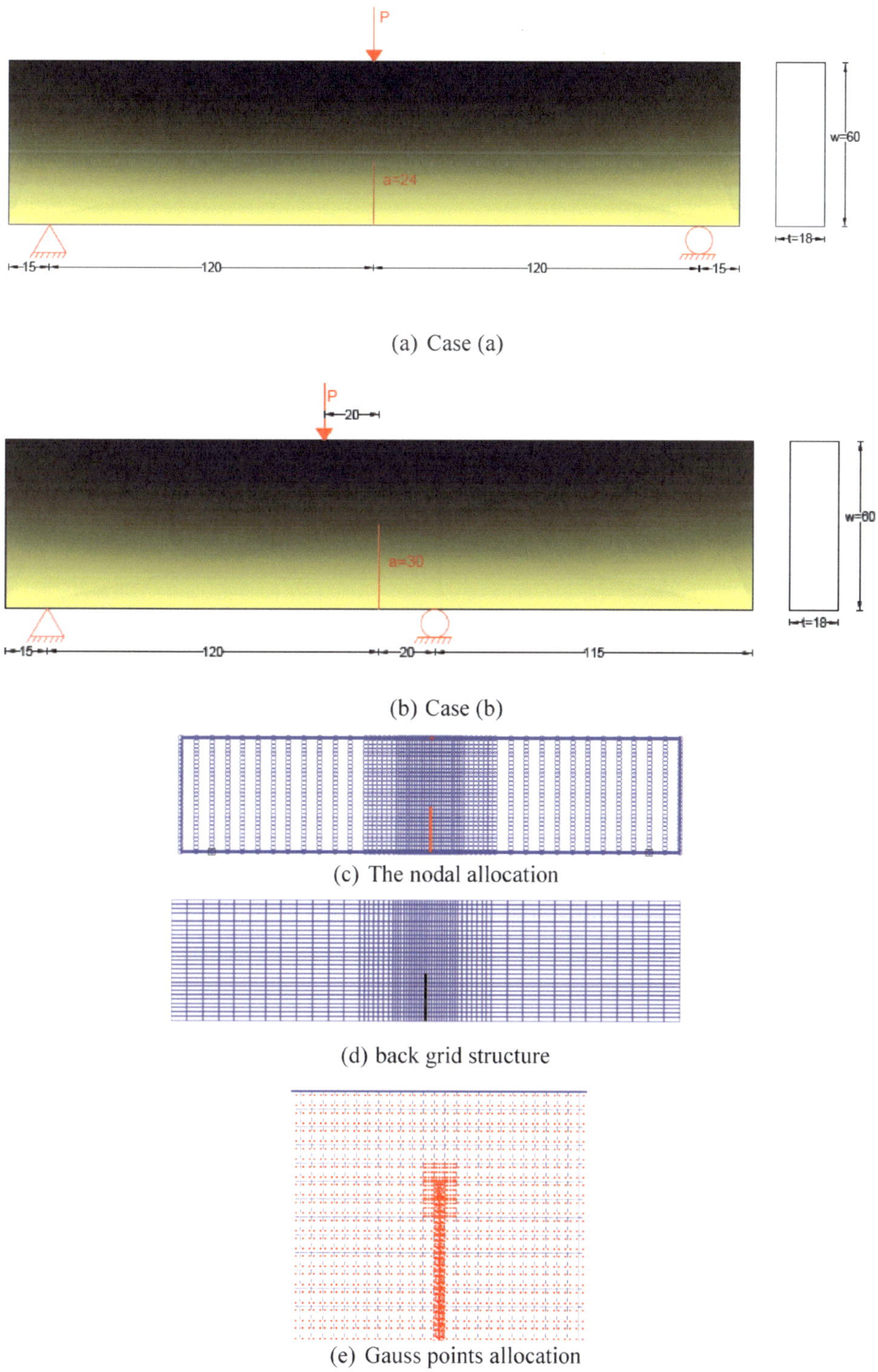

Figure 4.
Three-point bending cracked beam. (a) Case (a). (b) Case (b). (c) The nodal allocation. (d) back grid structure. (e) Gauss points allocation.

gradient beam. **Figures 8** and **9** depict results of KI and KII for the different relative size of the J-integral domain (rJ) respectively. The results of the proposed method remain accurate for a wide range of rJ values and the integral field size J (rJ) does not significantly influence the values of SIFs.

X_2	E (MPa)	ν	K_{cr} (Mpa $\sqrt{m}$)
0	1780	0.41	0.99
60	4000	0.39	1.19

Table 1.
Material characteristics of the graduated beam.

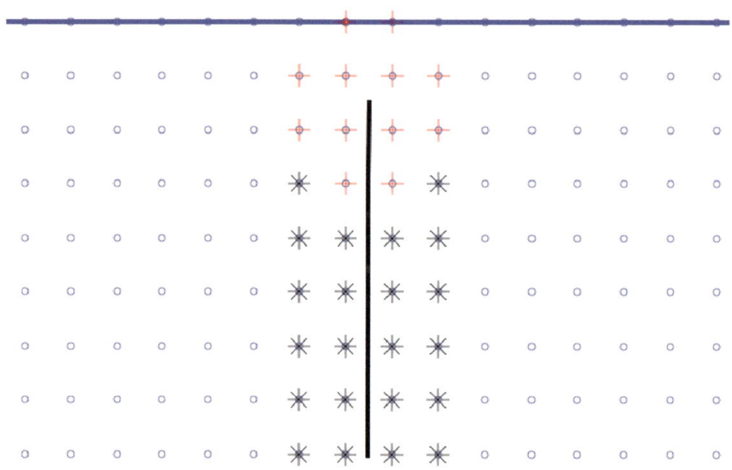

Figure 5.
Nodes of enrichment around the crack prior to the final step of crack propagation.

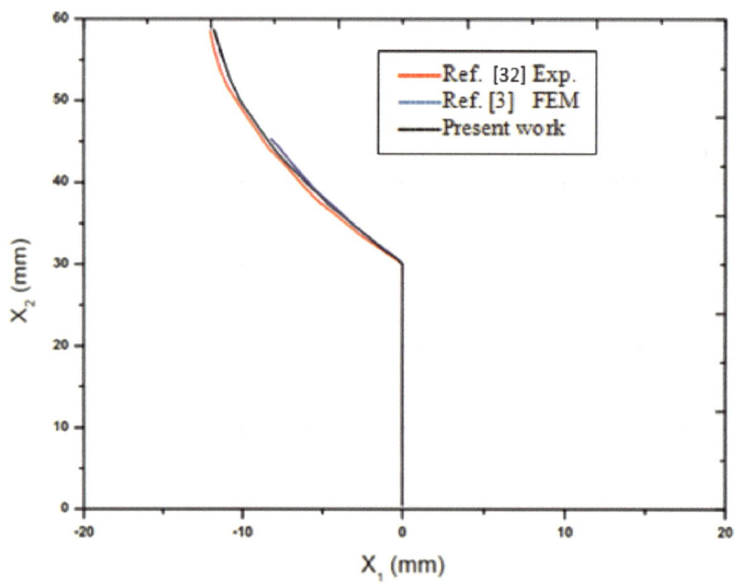

Figure 6.
Comparison of crack paths for a homogeneous beam (Case b).

5.2 Example 2

Figure 10 depicts the configuration and mechanical properties (**Table 2**) of the case study (2) that is bending four points with vertical cracks that perpendicular on the gradient of material.

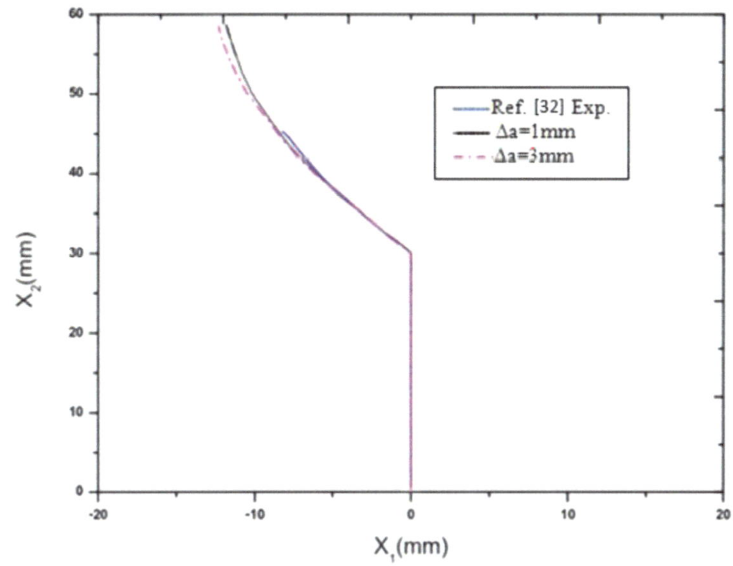

Figure 7.
The effect of increasing the crack length by current numerical simulations of case (b) on the crack path in a graded beam.

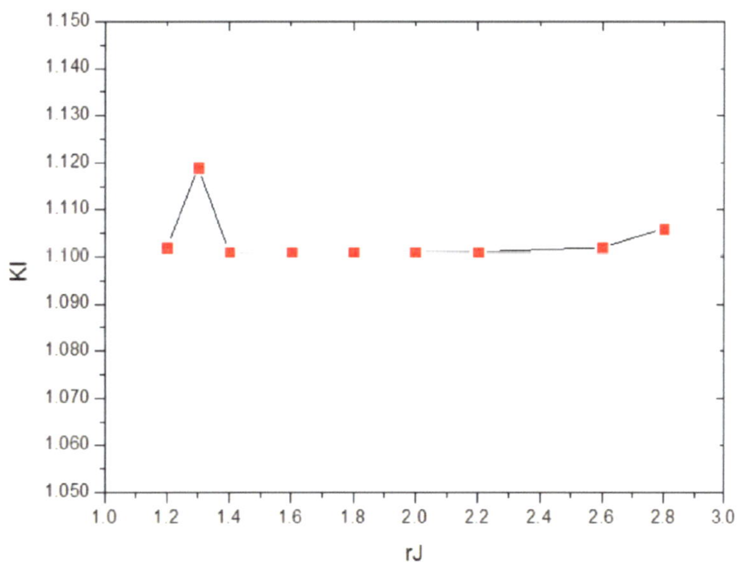

Figure 8.
K_I values for case (b) with different relative rJ.

Rousseau and Tippur [33] applied ξ that is zero on the left side of the stepping part, and one on the right side (**Figure 10**). In current work, A 64 x 28 back grid and 2070 non-uniform distribution nodes are adopted in this case (**Figures 11** and **12**), while more than 10,000 element and 30,000 nodes were adopted by [33] to study this case.

The results of the current research work give high accuracy with related references as depicted in **Tables 3** and **4**. The consistent of present research can be depicted in **Tables 3–5**. **Figure 13** gives a comparison of the effect of increasing the slit length on the slit path of the current work with experimental work [33] at

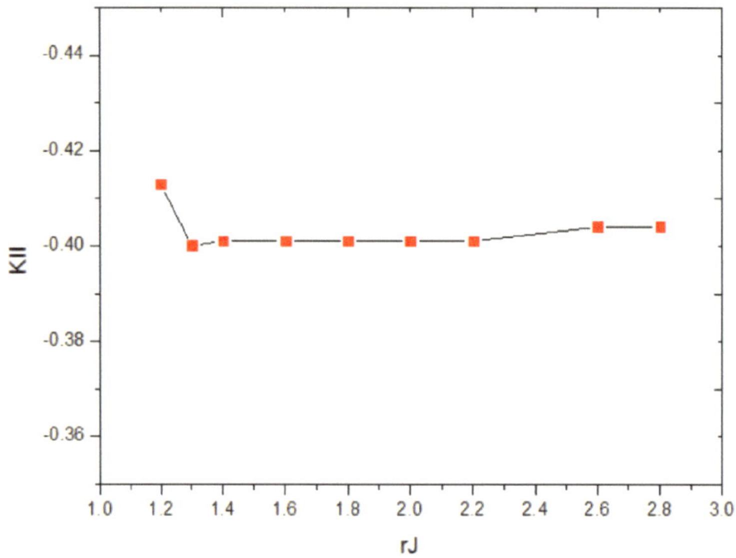

Figure 9.
K_{II} values for case (b) with different relative rJ.

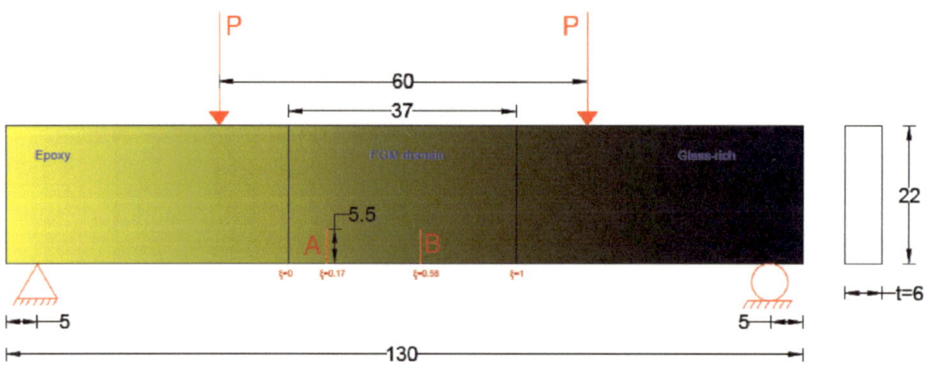

Figure 10.
Bending four points with vertical cracks on the physical gradient.

ξ	E (MPa)	ν	K_{cr} (Mpa $\sqrt{m}$)
0≤	3000	0.35	1.2
0.17	3300	0.34	2.1
0.33	5300	0.33	2.7
0.58	7300	0.31	2.7
0.83	8300	0.3	2.6
1≥	8600	0.29	2.6

Table 2.
Material characteristics of the graduated beam.

ξ = 0.17, 0.58, and ξ = 1.00. Finally, **Figures 14** and **15** appear the data of K_I and K_{II} for different relative rJ respectively. It is clear in this example that the growth of the crack is moving towards the soft side.

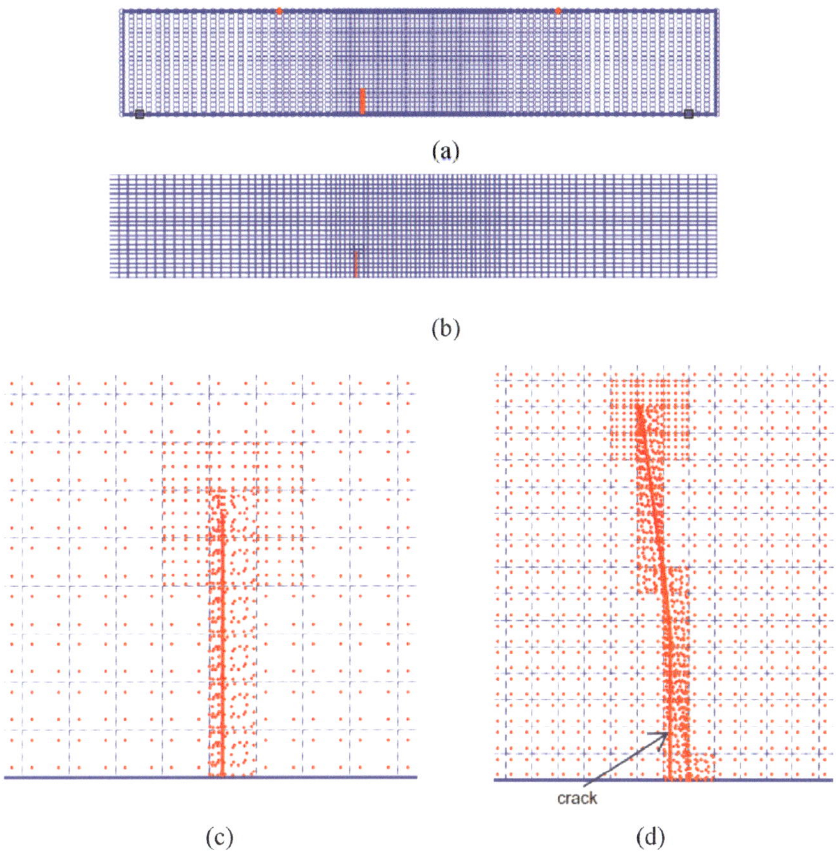

Figure 11.
(a) Distribution of 2070 irregular nodes, (b) Back grid structure, (c-d) Sub-triangulation technique upon initial fracture propagation.

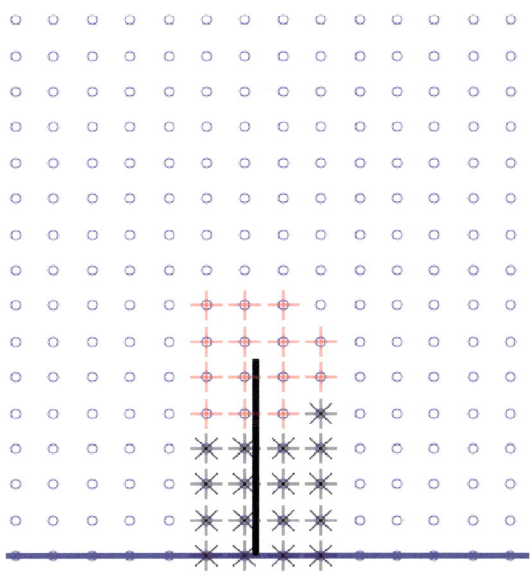

Figure 12.
Nodes of enrichment around the crack.

ξ	$K_I(Mpa\ \sqrt{m})$			$K_{II}(Mpa\ \sqrt{m})$		
	Present work	**Ref. [8]**	**Ref. [4]**	**Present work**	**Ref. [8]**	**Ref. [4]**
0.17	2.087	2.087	2.088	−0.116	−0.117	−0.127
0.58	2.689	2.694	2.695	−0.087	−0.085	−0.094
1	2.588	—	2.598	−0.010	—	−0.013

Table 3.
Stress intensity factors rapprochement at $d_{max} = 2$ *and* $r_J = 1.2$.

ξ	P_{cr} (N)			Crack initiation angle θ_0 (deg.)		
	Present work	**Ref. [8]**	**Ref. [4]**	**Present work**	**Ref. [8]**	**Ref. [4]**
0.17	255	250	249.3	6.901	7.009	7
0.58	298	300	298	4.020	3.998	4
1	295	—	289.9	0.601	—	0.5

Table 4.
Critical load of and crack initiation angle rapprochement at $d_{max} = 2$ *and* $r_J = 1.2$.

$\xi = 0.17; r_J = 1.2$	d_{max}		
	1.7	**2**	**2.3**
$K_I(Mpa\ \sqrt{m})$	2.081	2.087	2.091
$K_{II}(Mpa\ \sqrt{m})$	−0.112	−0.116	−0.111

Table 5.
SIFs for different support domain size.

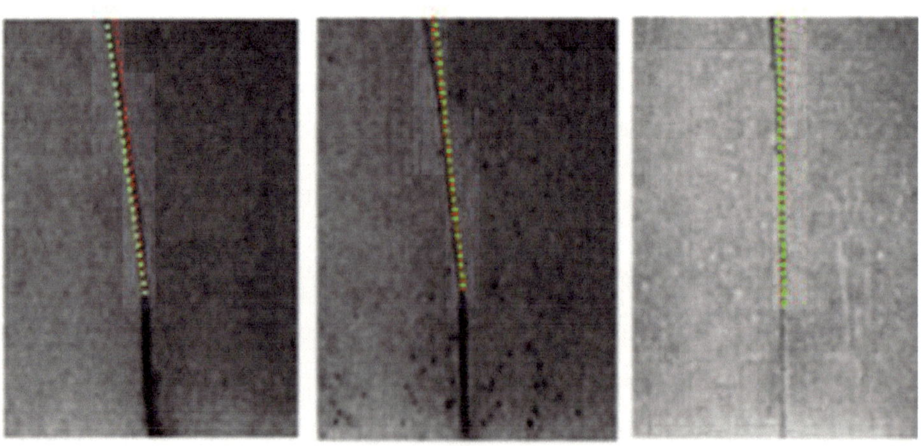

Figure 13.
Comparison of effect of increasing the crack length on the propagation path for current research and experimental work [33]. (Δa = 1 mm for red line, and Δa = 3 mm for green line).

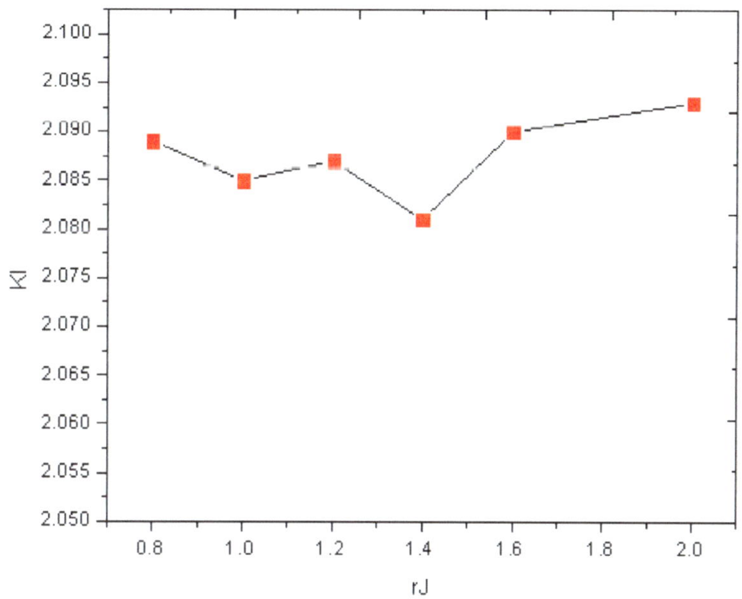

Figure 14.
KI values for case (2) vs. relative rJ.

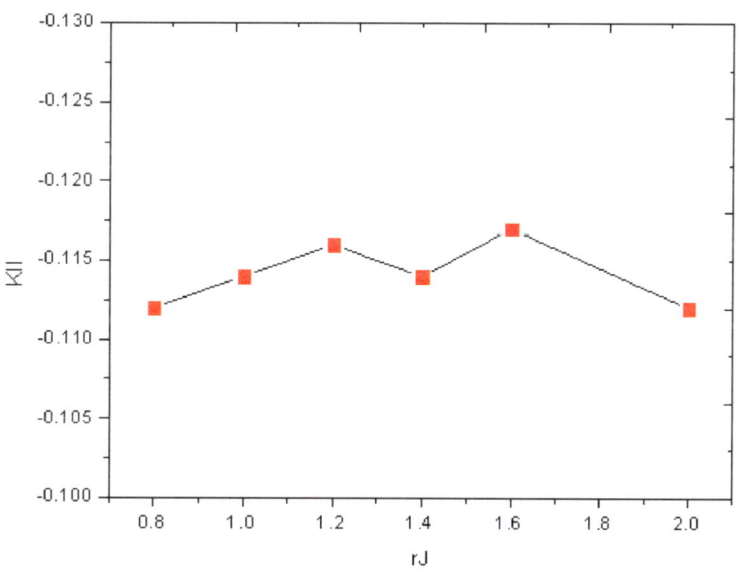

Figure 15.
K_{II} for case (2) vs. relative r_J.

6. Conclusion

The work mentioned in this work submitted the development of the XEFGM method to simulate crack propagation and compute stress intensity factors in mixed-mode fracture analysis of FGM beams under mixed mode and asymmetric loading. It has been demonstrated that XEFGM needs much less DOF than traditional FEM and XFEM to give the same accuracy levels. The adopt of sub-triangle technique for numerical integration, proper support field, and enrichment functions at the crack site has been shown to significantly increase the resolution of the solution. In addition, numerical simulation showed little effect of increasing incision length on the propagation path compared to the available experimental and numerical results. It is clear in this example that the growth of the crack is moving towards the soft side. There is no effect for increasing the crack length on the propagation path for current research with experimental work.

Additionally, the gradation of materials has no effect on the incision path of bending cases. Use of incompatible interaction integration method provides very accurate results for SIFs values.

Author details

Nathera A. Saleh* and Haider Khazal
Mechanical Engineering Department, Basrah University, Basrah, Iraq

*Address all correspondence to: nathera1971@yahoo.com

References

[1] R. M. Mahamood and E.T. Akinlabi, “Functionally Graded Materials”, springer, 1 Ed, ISBN 978-3-319-53755-9.

[2] B.N. Rao and S. Rahman, “Mesh-free analysis of cracks in isotropic functionally graded materials” Engineering Fracture Mechanics 70 (2003) 1–27.

[3] J.-H. Kim and G.H. Paulino, “Simulation of crack propagation in functionally graded materials under mixed-mode and non-proportional loading” International Journal of Mechanics and Materials in Design 1: 63–94, 2004.

[4] J.-H. Kim and G.H. Paulino, “On Fracture Criteria for Mixed-Mode Crack Propagation in Functionally Graded Materials”, Mechanics of Advanced Materials and Structures, 14:227–244, 2007.

[5] B.N. Rao and Sharif Rahman, “A continuum shape sensitivity method for fracture analysis of orthotropic functionally graded materials”, Mechanics of Materials 37 (2005) 1007–1025

[6] Rolf Mahnken, “Geometry update driven by material forces for simulation of brittle crack growth in functionally graded materials”, Int. J. Numer. Meth. Engng 2009; 77:1753–1788.

[7] Sayyed Shahram Ghorashi, Soheil Mohammadi, Saeed-Reza and Sabbagh-Yazdi, “Orthotropic enriched element free Galerkin method for fracture analysisof composites”, Engineering Fracture Mechanics Volume 78, Issue 9, June 2011, 1906-1927

[8] Hamid Bayesteh, Soheil Mohammadi, “XFEM fracture analysis of orthotropic functionally graded materials”, Composites: Part B 44 (2013) 8–25

[9] Haider Khazal, Hamid Bayesteh, Soheil Mohammadi, Sayyed Shahram Ghorashi, and Ameen Ahmed, “An extended element free Galerkin method for fracture analysis of functionally graded materials" mechanics of advanced materials and structures 23 (2016)513-528.

[10] N. Muthua,b,d, S.K.Maitib,n, B.G. Falzon c, WenyiYan, “Crack propagation in non-homogenous materials: Evaluation of mixed-mode SIFs, T-stress and kinking angle using a variant of EFG Method”, Engineering Analysis with Boundary Elements 72(2016)11–26.

[11] J.Li, J.Z.Liu, T.Korakianitis, P.H. Wen “Finite block method in fracture analysis with functionally graded materials”, Engineering Analysis with Boundary Elements, Volume 82, September 2017, Pages 57-67

[12] Popovich, V. A., et al. “Creep and thermomechanical fatigue of functionally graded Inconel 718 produced by additive manufacturing.” TMS annual meeting & exhibition. Springer, Cham, 2018.

[13] Zhao, S., et al. “Compressive and fatigue behavior of functionally graded Ti-6Al-4V meshes fabricated by electron beam melting.” Acta Materialia 150 (2018): 1-15.

[14] Farouq, Wafaa, Haider Khazal, and Abdul Kareem F. Hassan. “Fracture analysis of functionally graded material using digital image correlation technique and extended element-free Galerkin method.” Optics and Lasers in Engineering 121 (2019): 307-322.

[15] H. Khazal, A. Hassan, W. Farouq, and H. Bayesteh, “Computation of Fracture Parameters in Stepwise Functionally Graded Materials Using Digital Image Correlation Technique,” Materials Performance and Characterization 8, no. 1 (2019): 344-354.

[16] Abood, Ahmed M., Haider Khazal, and Abdulkareem F. Hasan. "Evaluation of mixed-mode stress intensity factor and T-stress in continuous epoxy glass functionally graded beam using digital image correlation." Materials Today: Proceedings (2021).

[17] Eischen JW. Fracture of nonhomogeneous materials. Int J Fract 1987; 34:3–22.

[18] Anderson TL. Fracture mechanics—fundamentals and applications. 2nd ed. Boca Raton, FL: CRC Press; 1995.

[19] P. Lancaster and K. Salkauskas "Surfaces Generated by Moving Least Squares Methods", Mathematics of Computation, Vol. 37, pp. 141-158, 1981

[20] G.R. Liu "Meshfree methods moving beyond the finite element method", second edition- 2010 by Taylor and Francis Group, LLC.

[21] S. Mohammadi, "Extended finite element method for fracture analysis of structures" First Edition 2007-Blackwell Publishers, UK.

[22] S. Mohammadi, "XFEM fracture analysis of composites", 2012, A John Wiley & Sons, Ltd. Publication.

[23] A. Asadpoure and S. Mohammadi, "Developing new enrichment functions for crack simulation in orthotropic media by the extended finite element method" Int. J. Numer. Meth. Engng 2007; 69:2150–2172.

[24] S.S. Ghorashi, S. Mohammadi and S. R.S.Yazdi, "Orthotropic enriched element free Galerkin method for fracture analysis of composites" Engineering Fracture Mechanics- Volume 78, Issue 9, June 2011, Pages 1906–1927.

[25] J. Dolbow "An extended finite element method with discontinuous enrichment for applied mechanics", Ph.D. thesis. Northwestern University, Evanston, IL, USA, 1999.

[26] J.-H. Kim and G. H. Paulino, "Consistent Formulations of the Interaction Integral Method for Fracture of Functionally Graded Materials" Journal of Applied Mechanics, ASME-May 2005,Vol. 72, pp. 351-364.

[27] S.S. Hosseini, H. Bayesteh and S. Mohammad," Thermo-mechanical XFEM crack propagation analysis of functionally graded materials", Materials Science and Engineering: A, Volume 561, 20 January 2013, Pages 285–302.

[28] L. Guo, F. Guo, H. Yu and L. Zhang, "An interaction energy integral method for nonhomogeneous materials with interfaces under thermal loading" International Journal of Solids and Structures 49 (2012) 355–365.

[29] J.-H. Kim and G.H. Paulino, "Mixed-mode fracture of orthotropic functionally graded materials using finite elements and the modified crack closure method", Engineering Fracture Mechanics, 69 (14–16), 1557–1586, 2002.

[30] J.-H. Kim and G. H. Paulino, "On fracture criteria for mixed-mode crack propagation in functionally graded materials", Mechanics of Advanced Materials and Structures, 14:227–244, 2007.

[31] Erdogan F, Sih GC. On the crack extension in plates under plane loading and transverse shear. ASME J Basic Eng 1963; 85(4):519–527.

[32] Galvez, J., Elices, M., Guinea, G.V. and Planas, J. (1996). Crack trajectories under mixed mode and non-proportional loading. International Journal of Fracture 81(2), 171–193.

[33] Rousseau C-E, Tippur HV. Compositionally graded materials with cracks normal to the elastic gradient. Acta Mater 2000; 48(16):4021–4033.

Chapter 6

Determining the Characteristics of Acoustic Emission in the Fatigue Crack Growth of Aluminum Alloy 2025 for Online Structural Monitoring

Javad Sharifi Ghaderi

Abstract

In the use of metals, due to industrial advances and the application of more dynamic loads, it is necessary to pay more attention to the fatigue issue. Non-destructive inspection methods are used to condition and health monitoring of structures at the time of production and even during the service life of parts. Among non-destructive methods, the acoustic emission method has become a standard and reliable method in recent years. In this project, the characteristics of acoustic emission in the fatigue crack growth of aluminum alloy 2025 for online structural monitoring have been investigated and determined. Acoustic emission tests have been performed in two parts: bending fatigue test with the aim of initiation of fatigue cracks in aluminum alloy 2025 specimens and following tensile tests with the aim of growth of fatigue cracks. The acoustic emission signals and parameters sent by the acoustic emission sensor during both tests were received and recorded by the acoustic emission software. According to the received acoustic emission information, various diagrams are plotted. Analyzing the results from online acoustic emission monitoring showed, the acoustic emission method can be considered as a suitable and reliable technique for detecting crack initiation and crack growth in aluminum alloy 2025.

Keywords: fatigue, bending fatigue test, tensile test, acoustic emission test, aluminum alloy 2025, acoustic emission signal, acoustic emission counts

1. Introduction

1.1 Fatigue failures

Fatigue failure is the most common type of failure because 75 to 90% of material failure in engineering components occurs due to cyclic loading [1]. This type of failure mainly occurs in systems where force or moments is applied continuously but vary in size. Failures that occur under dynamic loading conditions are called fatigue failures. There is no obvious change in the metal structure that fractures due to fatigue that it can be used as evidence to identify the causes of fatigue failure [1]. With the advance of industry and the increase in the number of machines such as cars, airplanes, compressors, pumps, turbines, etc. that are under repeated load and vibration, fatigue has become more common. The main reason that fatigue failure is dangerous is that it occurs suddenly and invisibly [1]. The beginning of the failure of the part due to fatigue is the initiation of microscopic cracks and then their growth. Crack growth continues until the cross-section of the metal is so small that the stress on it is greater than the ultimate strength of the metal so brittle failure occurs in the specimen. For this reason, failure will be sudden. The fatigue failure surface consists of a smooth area due to friction with the growth of cracks in the section and a rough area that is softly broken in the part when the load is intolerance by the section [2]. **Figure 1** shows fatigue failure of the Bonanza F33 propeller, which is made of aluminum alloy 2025.

Figure 1.
Fatigue failure of bonanza F33 propeller.

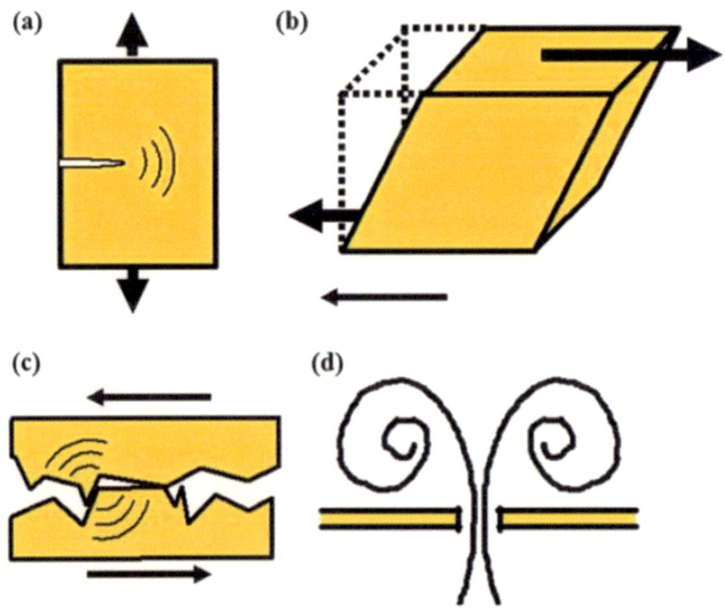

Figure 2.
Example of sources of acoustic emission. (a) Cracking. (b) Deformation and transformation. (c) Slip. (d) Leakage.

1.2 Acoustic emission

Non-destructive testing methods, especially acoustic emission methods, are used to condition monitor of engineering structures [3]. Acoustic emission as a phenomenon can be defined as transient elastic waves caused by internal micro-displacements in the materials of the tested structures. Acoustic emission, as defined by the American Society for Testing and Materials, refers to the class of phenomena whereby transient elastic waves are generated by the rapid release of energy from localized sources within material under stress [4]. Due to its high sensitivity, this method can detect processes such as micro-crack initiation and growth, displacement, failure, slip, leakage, or sediment separation [5]. **Figure 2** shows typical mechanisms that generate acoustic emission waves. The main sources of acoustic emission in metals are plastic deformation and crack growth processes, which are energy-release mechanisms at the scale of crystalline microstructure.

The acoustic emission method has advantages over other non-destructive testing methods, such as the dynamics of this inspection method and the ability to display crack growth online in the structure under load and during service. Other advantages of this method are high speed of testing, accurate location of defects, high efficiency, less sensitivity to the geometry of the part compared to other methods, and detection of very small and micro-scale defects [3].

2. Specimens and experimental equipment

2.1 Aluminum alloy 2025 specimens

According to the structure determination test, the material of the extracted specimen from the propeller blade was determined as aluminum alloy 2025. Due to the limited dimensions of the propeller blade and the impossibility of extracting the specimen with standard dimensions in the standard reference ASTM-E855–08, the dimensions of the test specimen are selected very close to the standard dimensions [6]. Based on this, the length, width, and thickness of the specimen equal to 160 mm, 30 mm, and 4 mm are selected and specimens with these dimensions are extracted. Also, to initiate a crack, a notch with a thickness of 1.5 mm and 1.25 mm width is created parallel to the width of the specimen and at a distance of 27 mm along the length of the part. This notch helps to increase the speed of crack initiation in the specimen. **Figure 3** shows the aluminum alloy 2025 specimen dimensions.

2.2 Bending fatigue machine

In bending fatigue tests with the aim of initiation of cracks from the notch created in the 2025 aluminum specimen, we need a bending fatigue machine. The fatigue machine designed to test aluminum alloy 2025 specimens are mounted on a lathe and receives the moment and force applied to the specimen for fatigue from the machine's engine. The rotational motion created by the motor is converted to linear motion through the crankshaft connected to the drill chuck of the device and the connecting rod. This linear motion is then transmitted to one side of the specimen by a bar. The other side of the sample is fixed with a clamp so that we can see the movement on only one side [7]. **Figure 4** shows the bending fatigue machine. The bending fatigue test is performed in the case of fixed grip loading with a rate of 12 mm per cycle.

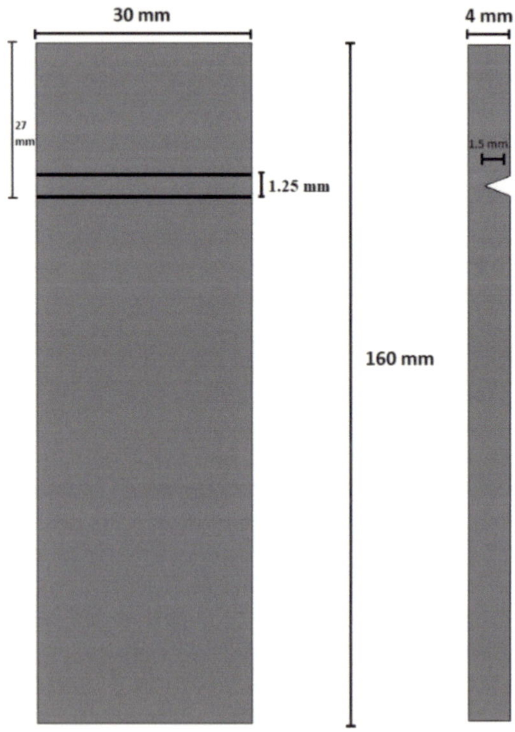

Figure 3.
Schematic and dimension of aluminum alloy 2025 specimen.

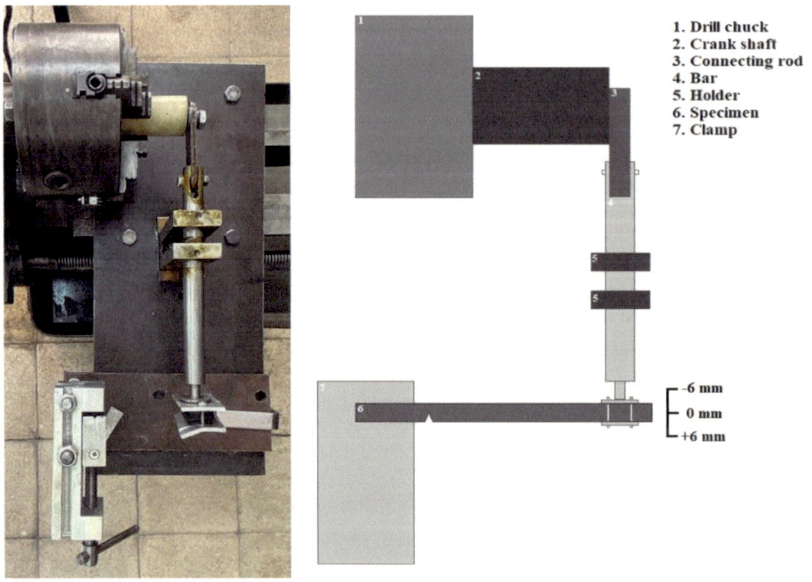

Figure 4.
Schematic and component of bending fatigue machine.

2.3 Tensile machine

In the tensile test, to record the signals emitted from the growth of fatigue cracks in aluminum alloy 2025, we need to grow the initiated cracks in the bending fatigue test by the tensile machine. This device is made by the HIWA company and has two jaws to connect the two sides of the specimen and a load cell to measure the tensile force. **Figure 5** shows the schematic of tensile machine. Cracked specimens are attached to both jaws of the machine on both sides, and after determining the tensile speed (1 mm/min in this test), the upper jaw of the machine starts to move upwards at the set speed and the initiated crack begins to grow [8].

2.4 Acoustic emission system

This system includes an acoustic emission sensor, a preamplifier, and acoustic emission software. The acoustic emission sensor used in this test is made of Lead-Zirconate with a diameter of 5 mm and a height of 4 mm of broad-band type and is connected to the preamplifier via a cable. The preamplifier has an input that can gain the signal received from the acoustic emission sensor with coefficients of 20, 40, and 60 dB. In this test, a coefficient of 40 dB was used and the output part sends the signal by cable to the computer for processing. The software installed on the computer is called AEwin for PCI-2, which allows us to set the parameters of acoustic emission testing, interpretation, display, and analysis of waveforms, adjust and display and compare several graphs, etc. [9]. **Figure 6** shows the schematic of acoustic emission system. After plotting various diagrams such as acoustic emission

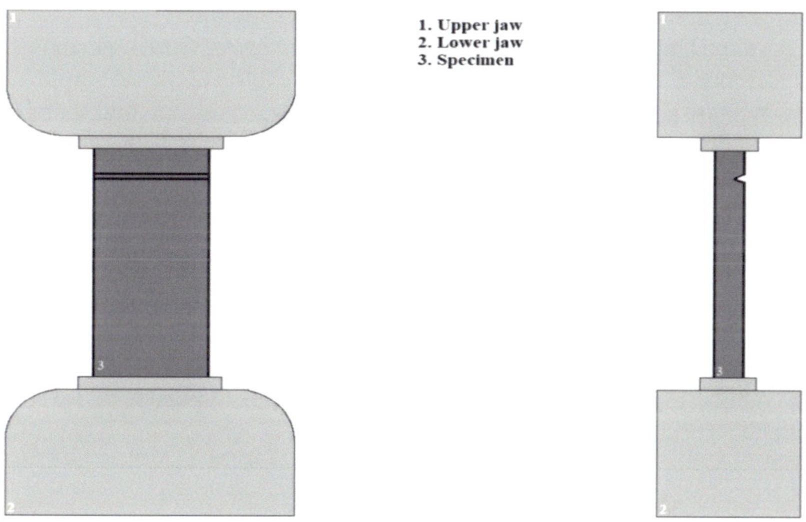

Figure 5.
Schematic and component of tensile machine.

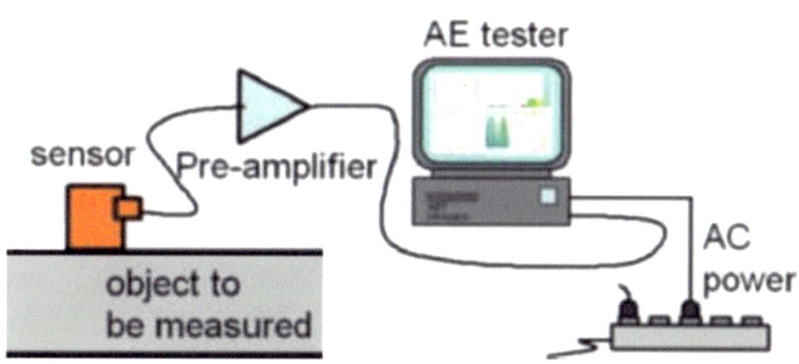

Figure 6.
Schematic of acoustic emission system.

signal amplitude vs. standard cycle diagram and acoustic emission cumulative count vs. standard cycle diagram by the acoustic emission system, and by analyzing the acoustic emission parameters such as signal amplitude, counts, rise time and duration time (In this test, acoustic emission amplitude and count were used), the crack initiation and crack growth can be determined.

The threshold was determined based on ambient noise in the bending fatigue test of 45 dB and in the tensile test of 20 dB, the frequency between 20 kHz to 1 MHz and the sampling rate of 2 MHz per second were determined in acoustic emission software.

3. Overview of experimental results

3.1 Bending fatigue test results (crack initiation)

3.1.1 Acoustic emission signal amplitude versus standard cycle diagram

The first plotted diagram is signal amplitude vs. normalized cycle. This diagram shows the signal amplitude in each standardized cycle. This diagram is important because it determines the recognizability of the signals and is also directly proportional to the magnitude of the event that occurred at the source [3, 10].

After examining the signal amplitude vs. standard cycles for all specimens, it was determined that in each specimen in a different cycle, the signal amplitude starts to increase sharply, then this amount reaches its maximum, and then begins to decrease. The amplitude of the start signal changes from 80–96% of the maximum signal amplitude in the specimens.

This increase in signal amplitude in the diagrams indicates the occurrence of an event within the test specimen. For example, the signal amplitude vs. normalized cycle diagram for specimen NO. 9, which is randomly selected from 9 specimens to explain in detail is shown in **Figure 7**. In the specified part, the signal amplitude in the standardized cycle 15 reaches 52 dB after the increase, then this value increases again until it reaches its maximum value in cycle 18, which is 54 dB, and then it starts to reduce. **Figure 8** shows acoustic emission signal amplitude vs. standard cycle for all 9 specimens.

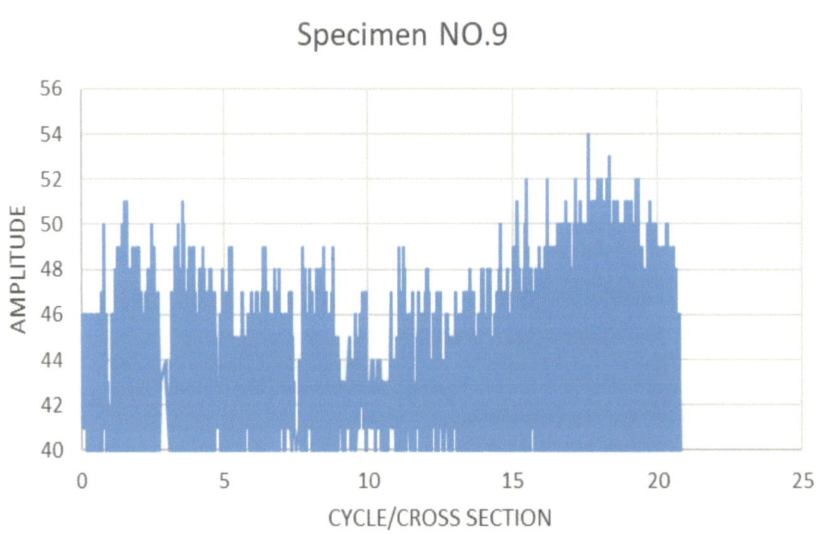

Figure 7.
Acoustic emission signal amplitude vs. standard cycle diagram of specimen NO. 9.

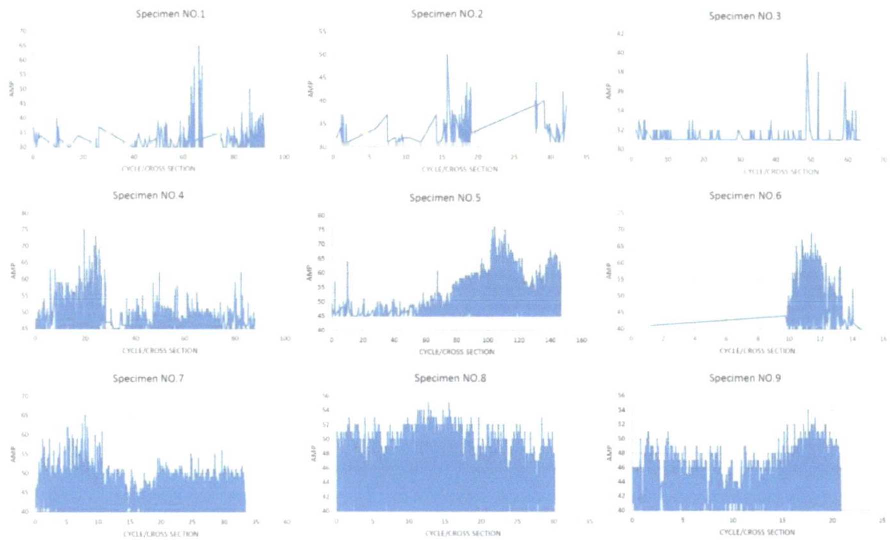

Figure 8.
Acoustic emission signal amplitude vs. standard cycle diagram for all 9 specimens.

Table 1 shows the maximum signal amplitude and amplitude which signal change start in all 9 specimens.

3.1.2 Acoustic emission cumulative count versus standard cycle diagram

The count is the number of pulses that exceed the specified threshold value [3, 10]. The cumulative count vs. normalized cycle diagram, in each standard cycle, shows the total number of counts of that cycle with previous cycles. The number of counts indicates the internal events of the material [11]. Therefore, where the counts reach their maximum value, the rate of internal events of materials is also at its maximum. After examining the amplitude vs. standard cycle diagram and determining the start cycle of changes in each specimen, to ensure the results are obtained, the cumulative count vs. standard cycle diagram is examined.

To study and analyze changes in cumulative graphs, slope changes in different parts of the graph are used. After examining the slope in different parts of the

Specimen number	Amplitude of start signal changes (dB)	Max. signal amplitude (dB)
NO. 1	52	65
NO. 2	40	50
NO. 3	33	40
NO. 4	63	75
NO. 5	68	76
NO. 6	65	69
NO. 7	62	65
NO. 8	53	55
NO. 9	52	54

Table 1.
Max. Signal amplitude and amplitude of start signal changes in each specimen.

cumulative count vs. standard cycle diagram, it was determined that before standard cycle NO. 10, an increase in slope is observed in all specimens, which is due to the instability of the conditions at the beginning of the test. After the simultaneous start of the acoustic emission system and the bending fatigue machine, the acoustic emission sensors receive the noise due to the mechanical vibration of the device after starting and display it as an acoustic emission signal which increases the slope in the cumulative counting vs. standard cycle diagram. The slope then continues almost uniformly until the same standard cycle as the amplitude of the signals began to increase, the slope of the cumulative count vs. standard cycle also begins to increase sharply. For example, the acoustic emission cumulative count vs. normalized cycle diagram for specimen NO. 9 is shown in **Figure 9**.

Figure 10 shows the cumulative count vs. standard cycle diagrams for specimens NO. 1 to 9. In diagram of specimen NO. 1, in the standard cycle, approximately 68 to 76 diagrams are in a horizontal line, which is the reason for stopping

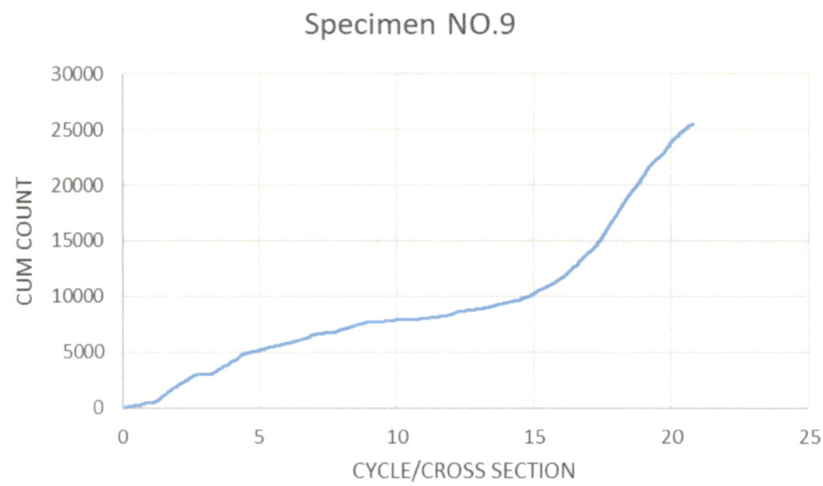

Figure 9.
Acoustic emission cumulative count vs. standard cycle diagram of specimen NO. 9.

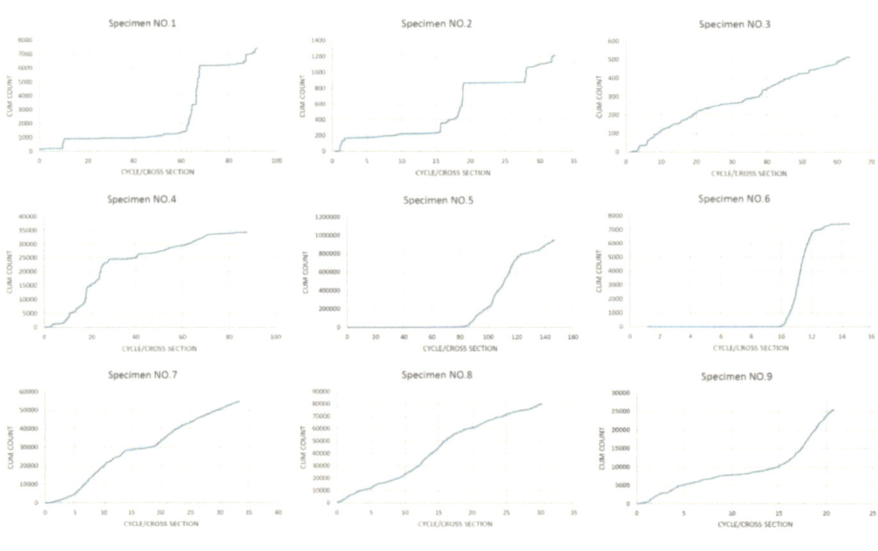

Figure 10.
Acoustic emission cumulative count vs. standard cycle diagram for all 9 specimens.

the test in 30 to 33 minutes. In diagram of specimen NO. 2, this event also occurred in the approximate cycle of 20 to 28, which is from 3 to 4:30 minutes.

For example, in specimen NO. 9, which was also examined in the signal amplitude vs. standard cycle diagram, in standard cycle NO. 15, which is the beginning of changes, it has a cumulative count of 10259, and in standard cycle NO. 20, which is the end of these changes, it has a cumulative count 23893. The slope of the change in the chart is 2870, which is 4 times more than before the start of the experiment when the slope is 703. In all specimens, this sharp increase in slope is noticeable. As mentioned, where the count reaches its maximum value, the rate of the internal events of the material is also at its maximum, in all specimens, the maximum value of the count is in the same range of changes. According to the above, examining the diagrams and the slope of different parts of the diagram, it can be concluded that crack initiation occurs when the slope of the diagram increases sharply. This increase in slope varies from 4 times to 16 times the slope before cracking between the tested specimens.

Table 2 shows maximum acoustic emission count and cycle which signal change start.

For better understand the simultaneity of increasing the signal amplitude and increasing the slope of the cumulative count diagram, the amplitude and cumulative count versus standard cycle diagrams plotted. **Figure 11** shows acoustic emission signal amplitude and cumulative count vs. standard cycle diagram for all 9 specimens.

3.2 Tensile test results (crack growth)

3.2.1 Aluminum alloy 2025 stress: Strain curve

The stress–strain curve is obtained by loading objects at a constant speed and measuring the amount of deformation in the tensile test. In this test, a specimen without any crack is installed on the tensile and is pulled at a speed of 1 mm per second. Using the results of this test, the stress–strain curve is plotted and the yield stress and ultimate stress in aluminum alloy 2025 are obtained. After testing and plotting the stress–strain curve, the ultimate stress level was 381.67 MPa and the yield stress was 275 MPa. The ultimate stress level for aluminum alloy 2025-T3 is 400 MPa in Ref. [12]. **Figure 12** shows stress–strain curve of aluminum alloy 2025 specimen that extracted from propeller blade. The difference between the measured stress and the reference stress is due to the life of the specimen used in the test because as the life of aluminum-containing copper alloy increases, this aluminum becomes brittle.

Specimen number	Cycle number of start signal changes	Max. acoustic emission count
NO. 1	64	523
NO. 2	15	111
NO. 3	39	13
NO. 4	8	436
NO. 5	102	396
NO. 6	11	70
NO. 7	6	48
NO. 8	11	67
NO. 9	15	39

Table 2.
Max. Acoustic emission count and cycle number of start signal changes in each specimen.

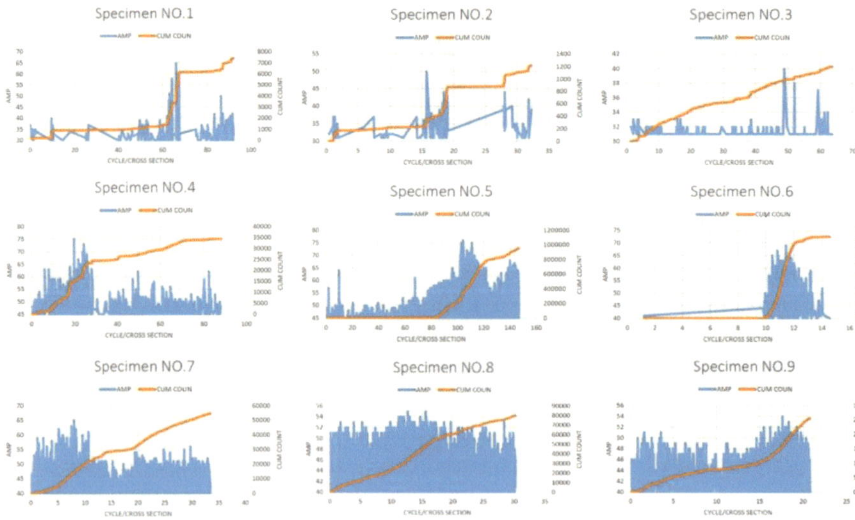

Figure 11.
Acoustic emission signal amplitude and cumulative count cumulative count vs. standard cycle diagram for all 9 specimens.

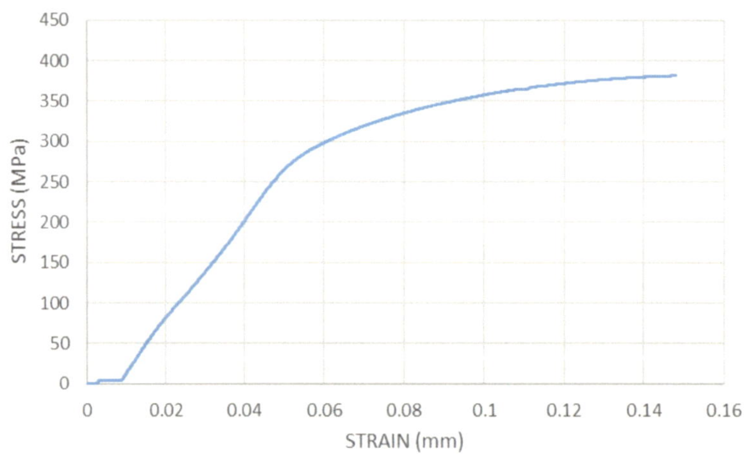

Figure 12.
Stress–strain curve of aluminum alloy 2025.

3.2.2 Acoustic emission count and stress versus time diagram

After performing the bending fatigue test and recording and analyzing the acoustic emission parameters due to crack initiation, it is necessary to subject the cracked specimens in the bending fatigue test to the tensile load to determine the parameters and characteristics of acoustic emission in fatigue crack growth. After performing the tensile test on 5 of the cracked specimens in the bending fatigue test, it is time to plot the count and stress vs. time diagram. This diagram shows the rate count and stress at each point in time of the test. Because all sources of additional signals and noise are blocked, the received signals are related to the acoustic emission activities inside the specimen. In general, these acoustic emission activities may be the result of plastic deformation or the growth of fatigue cracks created in the specimens. Because the test specimen is aluminum alloy 2025 with long life and brittle material and there is no sign of deformation in the specimen, the signals received by the sensor cannot be the plastic deformation signals so these signals are due to the growth of fatigue cracks.

After examining the count and stress vs. time diagram, it was found that with increasing stress, which is obtained by dividing the tensile force on the cross-section of the specimen, the count also increases. As mentioned in the previous paragraph, this increase in the count is related to the growth of cracks in the aluminum specimen. According to the **Figure 13**, which shows the count and stress vs. time diagram for specimen NO. 1, which is randomly selected from 5 specimens to explain in detail, the count does not increase continuously and the increase in count occurs after increasing the slope of the stress diagram.

Figure 14 shows the count and stress vs. time diagrams for all 9 specimens. As can be seen from the figure, in all specimens, the count increases sharply at the end of the test time, which indicates the highest crack growth activity during the test or an increase in crack growth rate with increasing force.

The highest increase of count for specimen NO. 1 occurred from 160 seconds to 167 seconds, where the highest rate of crack growth was observed. **Figure 15** shows the condition of the crack in 3 different times. Figure (a) shows the crack condition before the tensile test, when in the bending fatigue test the test is stopped immediately after observing the crack initiation. Figure (b) is after increasing the count

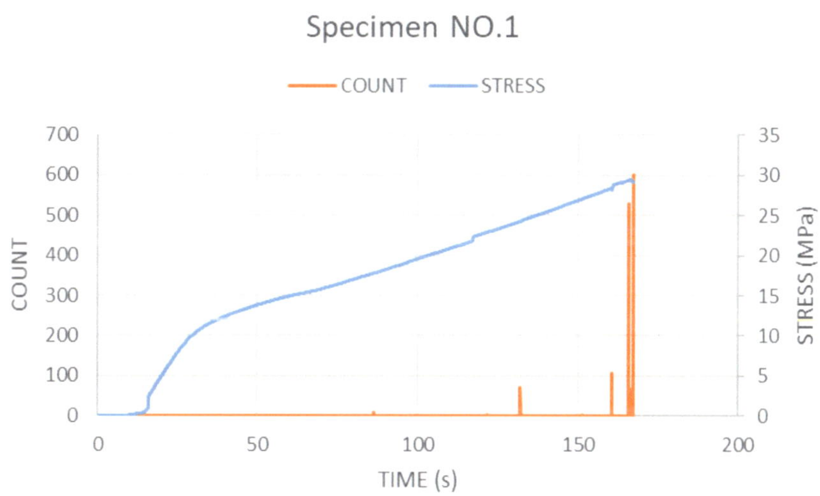

Figure 13.
Acoustic emission count, and stress vs. time diagram of specimen NO. 1.

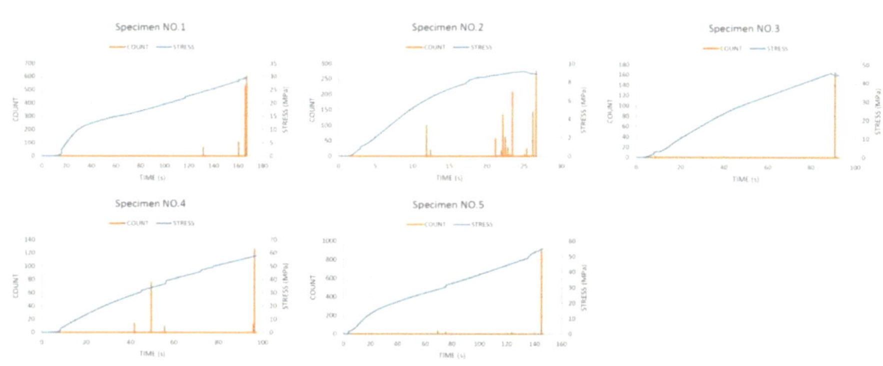

Figure 14.
Acoustic emission count, and stress vs. time diagram for all 9 specimens.

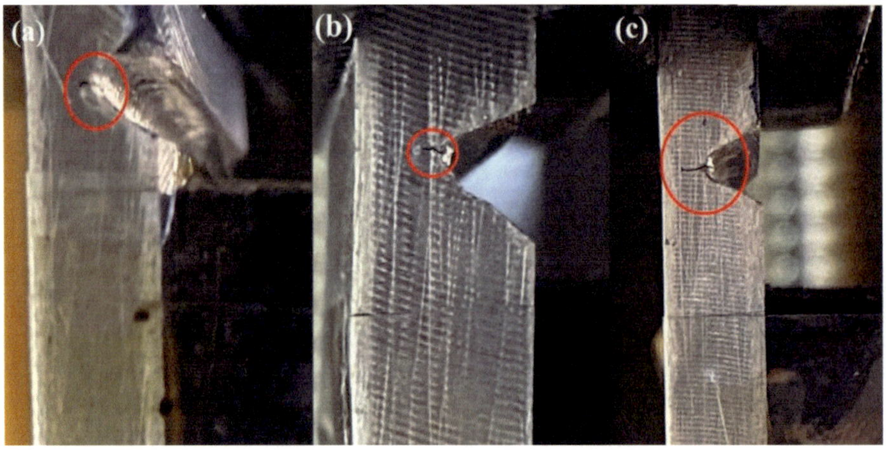

Figure 15.
Crack condition during specimen NO. 1 test (a) before tensile test (b) in 160 second (c) in 167 second.

at 160 seconds, where the count increases to 106 and the cumulative count to 189. Figure (c) also refers to a time of 167 seconds, where a sharp increase in the count, first at 165 seconds at 530 and then at 166 seconds at 602.

As the crack growth and the count diagram show, as time goes on and the stress and force increase, the count rate increase too, so the internal activity of the material and the crack growth increase, so that the maximum crack growth rate at the end of the test of each specimen.

4. Conclusion

Bending fatigue test and tensile test were performed on aluminum alloy 2025 specimens and acoustic emission characteristics were recorded and examined in each of the tests. One of the purposes of this project was to investigate the feasibility of the acoustic emission method in detecting the initiation and growth of fatigue crack growth in Bonanza f33 propellers. The analysis of this method was performed by examining specimens of propeller with aluminum alloy 2025.

The first part of the tests, bending fatigue test with the aim of crack initiation and recording the acoustic emission signals emitted from the aluminum alloy 2025 specimen. After reviewing and analyzing the amplitude vs. standard cycle diagram and cumulative count vs. standard cycle diagram, it was determined that the sharp increase of the signal amplitude to the maximum signal and the sharp increase of the slope of the cumulative count occurs in the same standard cycle. This time can be attributed to crack initiation. The slope of the count vs. standard cycle at the start of crack initiation increases more than four times this slope from the start of the test.

The second part of the tests, the tensile test was performed to grow the cracks created in the previous stage and record the acoustic emission signals caused by the growth of fatigue cracks in aluminum alloy 2025. To determine the condition of the specimens and plot a stress–strain curve, first the tensile test was performed on a specimen of aluminum alloy 2025 without a notch. After plotting the stress–strain curve, the yield stress was equal to 275 MPa and the ultimate stress was equal to 381.67 MPa.

Analysis of the parameters and diagrams of count and stress vs. time in the tensile test shows an increase in the count with increasing force. The highest rate of crack growth occurs at the end of each test, because the highest number of counts, which indicates the internal events of a specimen, is observed at the end of the test. An increase in the number of counts in the loading phase indicates the growth of cracks.

As a final conclusion, despite the brittle material of aluminum alloy 2025, the acoustic emission method is a reliable, accurate, and high-efficiency method to identify the initiation and growth of fatigue crack in this aluminum alloy.

Author details

Javad Sharifi Ghaderi
Shahid Sattari University of Aeronautical Engineering, Tehran, Iran

*Address all correspondence to: javadsharifighaderi@gmail.com

References

[1] Chawla K. K, Meyers M.A. Mechanical behavior of materials. 2nd ed. Cambridge university press; 2008. 1132 p. DOI:10.1017/CBO9780511810947

[2] Kushan M. C, Diltemiz S. F, Sackesen I, Engineering failure analysis: Failure analysis of an aircraft propeller. FEBS Letters. 2007;513:1693-1700. DOI:10.1016/j.engfailanal.2006.11.069

[3] Mizutani Y. Practical acoustic emission testing. 2nd ed. Springer; 2006. 1132 p. DOI:10.1007/9784431550723

[4] Nondestructive evaluation and test. ASTM hand book. 9th ed.; 1994. Vol 17. 1132 p. DOI:10.1520/STP624-EB

[5] Peng-jing Zhao, Yan-dond Sun, Jing-pin Jiao, Gang Fang, Engineering fracture mechanics: Correlation between acoustic emission detection and microstructural characterization for damage evolution. Vol. 230, May 2020. DOI:10.1016/j.engfracmech.2020.106967

[6] Bending Test Methods for Bend Testing of Metallic Flat Materials for Spring Application Involving. ASTM; 2013. 1132 p. DOI:10.1520/E0855-08R13

[7] Pritam. V. Kulkarni, P. J. Sawant, V. V. Kulkarni, Materials today: Design and development of plane bending fatigue testing machine for composite material. 2018, 11563-11568. DOI:10.1016/j.matpr.2018.02.124

[8] Standard test methods for tension testing of metallic materials E8/E8M. An American National Standard; 2013. DOI:10.1520/E0008_E0008M-16A

[9] Zinoviy Nazarchuk, Valentyn Skalskyi, Oleh Serhiyenko. Acoustic emission methodology and application. 1st ed. Springer; 2017. DOI:10.1007/978-319-49350-3

[10] Prosser W.H. The Propagation Characteristics of the Plate Modes of Acoustic Emission Waves in Thin Aluminum Plates and Graphite/Epoxy Composite Plate and Tubes [Thesis]. Maryland: John Hopkins University; 1991.

[11] Blanchette Y, Dickson J. I, Bassim M. N, Engineering fracture mechanics: Acoustic emission behavior crack growth of 7075-T651 Al alloy. FEBS Letters. 1986;513:647-656. DOI: 10.1016/0013-7944(86)90240-7

[12] AZoM. Aluminum 2025 alloy (UNSA92025) [Internet]. 2013. Available from: www.azom.com/article.aspx?ArticleD=8718 [Accessed: 2013-05-17]